Foundations of the
**Probabilistic Mechanics
of Discrete Media**

FOUNDATIONS & PHILOSOPHY OF SCIENCE & TECHNOLOGY SERIES

General Editor: MARIO BUNGE, *McGill University, Montreal, Canada*

Some Titles in the Series

AGASSI, J.
The Philosophy of Technology

ALCOCK, J.
Parapsychology: Science or Magic?

ANGEL, R.
Relativity: The Theory and its Philosophy

BUCHTEL, H. A.
The Conceptual Nervous System

BUNGE, M.
The Mind–Body Problem

GIEDYMIN, J.
Science and Convention

HATCHER, W.
The Logical Foundations of Mathematics

SIMPSON, G.
Why and How: Some Problems and Methods in Historical Biology

WILDER, R. L.
Mathematics as a Cultural System

Pergamon Journals of Related Interest

STUDIES IN HISTORY AND PHILOSOPHY OF SCIENCE*

Editor:
Professor Gerd Buchdahl, Department of History and Philosophy of Science, University of Cambridge, England

This journal is designed to encourage complementary approaches to history of science and philosophy of science. Developments in history and philosophy of science have amply illustrated that philosophical discussion requires reference to its historical dimensions and relevant discussions of historical issues can obviously not proceed very far without consideration of critical problems in philosophy. STUDIES publishes detailed philosophical analyses of material in history of the philosophy of science, in methods of historiography and also in philosophy of science treated in developmental dimensions.

*Free specimen copy available on request

Foundations of the Probabilistic Mechanics of Discrete Media

by

D. R. AXELRAD
Micromechanics Research Laboratory
McGill University

PERGAMON PRESS
OXFORD · NEW YORK · TORONTO · SYDNEY · PARIS · FRANKFURT

U.K.	Pergamon Press Ltd., Headington Hill Hall, Oxford OX3 0BW, England
U.S.A.	Pergamon Press Inc., Maxwell House, Fairview Park, Elmsford, New York 10523, U.S.A.
CANADA	Pergamon Press Canada Ltd., Suite 104, 150 Consumers Rd., Willowdale, Ontario M2J 1P9, Canada
AUSTRALIA	Pergamon Press (Aust.) Pty. Ltd., P.O. Box 544, Potts Point, N.S.W. 2011, Australia
FRANCE	Pergamon Press SARL, 24 rue des Ecoles, 75240 Paris, Cedex 05, France
FEDERAL REPUBLIC OF GERMANY	Pergamon Press GmbH, Hammerweg 6, D-6242 Kronberg-Taunus, Federal Republic of Germany

Copyright © 1984 D. R. Axelrad

All Rights Reserved. No part of this publication may be reproduced, stored in a retrieval system or transmitted in any form or by any means: electronic, electrostatic, magnetic tape, mechanical, photocopying, recording or otherwise, without permission in writing from the publishers.

First edition 1984

Library of Congress Cataloging in Publication Data
Axelrad, D. R.
Foundations of the probabilistic mechanics of discrete media.
(Foundations & philosophy of science & technology series)
Bibliography: p.
Includes index.
1. Mechanics. 2. Markov processes. 3. Random fields. I. Title. II. Series: Foundations & philosophy of science & technology.
QA808.A93 1983   531   83-8080

British Library Cataloguing in Publication Data
Axelrad, D. R.
Foundations of the probabilistic mechanics of discrete media.—(Foundations & philosophy of science & technology)
1. Probabilities
I. Title II. Series
519.2'6   QA273.43

ISBN 0-08-025234-6

Printed and bound in Great Britain by
William Clowes Limited, Beccles and London

Preface

THE primary objectives of this monograph are twofold: to provide an account of probabilistic functional analysis and to show its applicability in the formulation of the behaviour of discrete media including microstructural effects. Due to the inherent random nature of the physical and configurational characteristics of discrete media, a formulation based on microscopic properties alone cannot be brought into line with the conventional deterministic macroscopic relations.

Although quantum mechanics has long been recognized as a stochastic theory, the introduction of probabilistic concepts and principles to classical mechanics has in general not been attempted. In the present study the view is taken from the onset, that the significant field quantities of a discrete medium are random variables or functions of such variables. Hence, the probabilistic mechanics of discrete media is based on the mathematical theory of probability and the axiomatics of measure theory. In developing a general theory for the behaviour of discrete media it becomes necessary to consider the evolution of the field variables with time and the associated probability measures. It is convenient for this purpose to employ the notion of an abstract dynamical system, which represents the motion of the medium in the physical space.

Most of the mathematical models of discrete media can be obtained by the application of the theory of Markov processes and Markov random fields. For the sake of brevity the emphasis in the presentation is on general results rather than on detailed examples. A brief review of functional analysis is given in the first chapter of the text. The significant theorems required in the subsequent analysis are given without proof, since the latter are readily available in the texts cited in the bibliography of this volume. Markov processes and Markov random fields are dealt with in Chapter 2. The fundamental concepts and axiomatic definitions of field variables employed in probabilistic mechanics are given in Chapter 3, which is also concerned with the general formulation of probabilistic mechanics. In Chapter 4 the mechanics of structured solids is discussed and a general probabilistic deformation theory is developed. The stability of microstructures in solids is briefly treated. Chapter 5 deals with the molecular dynamics of simple fluids and the formulation concerning such fluids on the basis of probabilistic concepts.

The author acknowledges the Natural Sciences and Engineering Research

Council of Canada, and the Faculty of Graduate Studies and Research of McGill University, without whose generous support the work in the Micromechanics Research Laboratory would not have been possible.

The author would like to express his sincere thanks to Professor Mario Bunge for the encouragement and support given to him during the preparation of this monograph. It is also a pleasure to acknowledge Professor O. Mahrenholtz, Director of the Institute of Mechanics, University of Hanover, and Professor Th. Lehmann of the Institute of Mechanics, Ruhr University, Bochum, for their kind invitations at various times to deliver seminars and courses concerning probabilistic mechanics at their Institutes.

The author would like to extend his thanks to Drs. M. Ostoja-Starzewski and K. Rezai for their conscientious assistance during the preparation of the text. Special thanks are due to Mr. J. Turski for his assistance in proofreading of the entire manuscript. Finally, the author expresses his thanks to Mrs. M. L. Powell for so patiently typing the original manuscript and the clerical work associated with the preparation of this volume.

Montreal, January 1983 D. R. AXELRAD

Contents

1. Mathematical Preliminaries   1

1.1   Introduction   *1*
1.2   Basic concepts and definitions of functional analysis   *1*
    1.2.1   Set theory   *2*
    1.2.2   Topological and measurable spaces   *5*
    1.2.3   Measures, distributions in topological spaces   *11*
    1.2.4   Linear operators and semi-groups   *14*
1.3   Probability, random variables and processes   *18*
    1.3.1   Probability   *19*
    1.3.2   Random variables and their distributions   *23*
1.4   Basic stochastic processes   *33*
1.5   Random fields   *42*

2. Markov Processes and Markov Random Fields   46

2.1   Introduction   *46*
2.2   Markov theory and dynamical semi-groups   *47*
2.3   Markov random fields   *57*

3. General Formulation of Probabilistic Mechanics   59

3.1   Introduction   *59*
3.2   States of elements of the structure   *60*
3.3   Fundamental concepts and definitions   *62*
3.4   The structure of the probabilistic function space $\mathscr{X}$ *66*
3.5   Interaction effects, potentials   *70*
3.6   Duality of spaces   *75*
3.7   Operational forms of the macroscopic constitutive relations   *79*

4. Probabilistic Mechanics of Solids   82

4.1   Introduction   *82*
4.2   Elements of the microstructure in solids   *82*
4.3   Interaction effects   *84*
4.4   General probabilistic deformation theory   *93*
4.5   Stability of microstructures   *102*
4.6   Material operators and boundary-value problems   *110*
4.7   Dynamics of structured solids   *114*

5. Probabilistic Mechanics of Fluids   121

5.1   Introduction   *121*
5.2   Molecular dynamics of simple fluids   *122*

5.3	*Response and fluctuation theory*	*132*
5.4	*Probabilistic mechanics of discrete fluids*	*142*
5.5	*Markov theory in the mechanics of discrete fluids*	*148*

Bibliography   **156**

Subject index   **163**

1
Mathematical Preliminaries

1.1 INTRODUCTION

Although the history of probability theory goes back as far as the seventeenth century, it was not developed on a strict mathematical basis until the end of the nineteenth century. D. Hilbert recognized the need for an axiomatic foundation of the theory of probability, but a satisfactory foundation was not achieved until the introduction of Kolmogorov's theory of probability spaces [1, 2]. The notion of a probability space is fundamental in the theory of Markov processes and their application to probabilistic mechanics. The basic concepts of probability are: a random variable, an event and the probability of an event. A given set of events forms a Boolean algebra that is defined with respect to the operations on the set. In this sense probability may also be regarded as a measure on the Boolean algebra of events, which is finitely additive and positive. Hence considerations of events and their probabilities become synonymous with the study of measures on fields of sets. However, not all sets or subsets of a topological space are suitable to form a measure. In dealing with the probabilistic mechanics of structured media two distinct features of the latter must be recognized. First, there exists a multitude of singular surfaces (interfaces) within a finite volume of the material body and second, the elements of the structure of actual materials exhibit random geometric and physical characteristics.

In view of these facts the probabilistic mechanics approach considers the relevant field quantities as random variables or functions of such variables together with their corresponding distribution functions, which form a set of measures. Hence the formal structure of the present theory is based on the mathematical theory of probability and the axiomatics of measure theory.

1.2 BASIC CONCEPTS AND DEFINITIONS OF FUNCTIONAL ANALYSIS

In order to establish the mathematical foundations of probabilistic mechanics certain concepts of topology, functional analysis and the theory of probability are required. It is the purpose of this and subsequent sections of this chapter to give certain definitions and theorems concerning sets, algebras, measures, topological spaces, etc. For a more comprehensive study and the proofs of various theorems cited in these sections, the reader is referred to the texts listed in the bibliography of this volume.

1.2.1 Set theory

(i) *Sets (definitions)*

A set is a collection of elements. The notation $x \in A$ means that x is an element of A. If x is not an element of A, it is written as $x \notin A$. An empty set or null set is denoted by ϕ. It does not contain any elements. If every element of a set A_1, is also an element of a set A_2, then A_1 is a subset of A_2: $A_1 \subseteq A_2$ or $A_2 \supseteq A_1$ or briefly $A_1 \subset A_2$, $A_2 \supset A_1$. The sets A_1 and A_2 are equal, if $A_1 \subseteq A_2$ and $A_2 \subseteq A_1$. The union of the sets A_1 and A_2 is denoted by $A_1 \cup A_2$. It is the set of all elements x that belong to either A_1 or to A_2.

The union of an arbitrary number of sets A_n or $\bigcup_n A_n$ consists of all elements x, which are contained in at least one of the sets A_n. The intersection of sets A_1 and A_2 is denoted by $A_1 \cap A_2$. It is the set of all elements which belong both to A_1 and A_2. The intersection of an arbitrary number of sets A_n or $\bigcap_n A_n$ consists of all elements that belong to every set A_n. One calls two sets A_1 and A_2 disjoint, if their intersection is void, i.e. if $A_1 \cap A_2 = \phi$.

If a sequence of sets $\{A_n\}$, $n = 1, 2, \ldots$ is a disjoint family of sets the union $\bigcup_n^\infty A_n$ can also be written as $\sum_{n=1}^\infty A_n$.

The subset of a set A consisting of all elements $x \in A$ with a certain property p is denoted by $\{x; p\}$. A proper subset A_1 of the set A is one which is not equal to A written as $A_1 \subset A$, $A_1 \neq A$ or briefly $A_1 \nsubseteq A$. If A is a subset of X, the complement of A relative to X consists of all elements $x \in X$ that are not contained in A. It is denoted by A' or $X \setminus A$. If A_1 and A_2 are two subsets of a set A, the difference $A_1 \setminus A_2$ is the totality of elements $x \in A$, which is not contained in A_2, i.e. $A_1 \setminus A_2 = A_1 \cap A_2'$. The symmetric difference of A_1 and A_2 or $A_1 \Delta A_2$ means $(A_1 \setminus A_2) \cup (A_2 \setminus A_1)$.

The Cartesian product $X \times Y$ of the sets X, Y is the set of all ordered pairs (x, y) in which $x \in X$ and $y \in Y$.

The partition of the set X is the class $\{X_n\}$ of non-empty disjoint subsets of X, whose union is X and where $n \in I$ is a positive integer of the index set I.

(ii) *Mappings*

The terms mapping, function and transformation are synonymous in analysis. Thus the notation $f: X \to Y$ means that if f is a single valued function

1.2 Basic concepts and definitions of functional analysis

whose domain $\mathcal{D}(f)$ is X, its range $\mathcal{R}(f)$ is contained in Y. A mapping of X into Y is also called a function on X with values in Y. The mapping f from X into Y associates with every element $x \in X$ a uniquely determined element $y \in Y$. For any two mappings $f: X \to Y$ and $g: Y \to Z$ one can define a composite mapping by $gf: X \to Z$, which is the mapping $g \circ f: X \to Z$ by $x \mapsto g(f(x))$. One can symbolically write $f(M)$ to denote the subset $\{f(x); x \in M\}$ of Y, where $f(M)$ is referred to as the image of M under the mapping f.

The notation of $f^{-1}(N)$ designates the subset $\{x; f(x) \in N\}$ of X, where $f^{-1}(N)$ is called the inverse image of N under the mapping f. If the mapping $f: X \to Y$ is such that for every $y \in \mathcal{R}(f)$ there is only one $x \in X$ with $f(x) = y$, then the mapping f has an inverse f^{-1} or the mapping is one–one (injective). Hence, if f has an inverse, it follows that:

$$\left. \begin{array}{l} f^{-1}(f(x)) = x \quad \text{for} \quad \forall x \in \mathcal{D}(f) \\ f(f^{-1}(y)) = y \quad \text{for} \quad \forall y \in \mathcal{R}(f) \end{array} \right\} \quad (1.1)$$

The function f maps X onto Y (surjective), if $f(X) = Y$ and into Y, if $f(X) \subseteq Y$. The mapping f is bijective, if it is one–one and onto (see also [3, 4, 5, 6]).

(iii) *Set relations*

A relation between two sets X and Y is a subset R of $X \times Y$. The elements x and y of these sets are R-related, if $(x, y) \in R$. An equivalence relation in X is written as $R \subset X \times X$ and has the following properties:

$$\left. \begin{array}{l} \text{(i) } (x, x) \in R \quad \text{for} \quad \forall x \in X \\ \text{(ii) } (x, y) \in R \Rightarrow (y, x) \in R \quad \text{for} \quad \forall x, y \in X \\ \text{(iii) } (x, y) \in R \quad \text{and} \quad (y, z) \in R \Rightarrow (x, z) \in R \\ \quad \text{for} \quad \forall x, y, z \in X \end{array} \right\} \quad (1.2)$$

An equivalence relation between x and y is written as $x \sim y$. The equivalence class of $x \in X$ is the subset $\{y : y \sim x\}$ of X. X is the disjoint union of equivalence classes. The set of these classes is denoted by X/R (not to be confused with $X \setminus R$). A partial ordering in X is the relation $R \subseteq X \times X$ satisfying the properties (i, iii) in (1.2) and (ii)′ $(x, y) \in R$ and $(y, x) \in R \Rightarrow x = y$. It is usually denoted $x \lesssim y$ or x is related to y by the partial ordering $\lesssim$. The elements $x, y, z \in P$ of a partially ordered set P, if $x \lesssim z$ and $y \lesssim z$, where z is the upper bound for x and y. If $z \lesssim t$ for all $t \in P$ that are upper bounds for x, y then z is the least upper bound or supremum of x, y denoted by $z = \sup(x, y)$. Analogously a lower bound and greatest lower bound or infimum of x, y is written as $\inf(x, y)$. A partially ordered set is said to be linearly ordered, if for any two elements $x, y \in P$ either $x \lesssim y$ or $y \lesssim x$.

An internal operation (binary operation) on a set A is the mapping from $A \times A$ into A. For two sets A_1, A_2 the mapping from $A_1 \times A_2$ into A_2 is called an external operation on A_2.

4   Mathematical Preliminaries

A group is a set X together with an internal operation such that:

(i) $(x\,y)z = x(y\,z)$ for all $x, y, z \in X$,
(ii) an element $e \in X$ called the identity, where
$$xe = ex = x \quad \text{for} \quad \forall x \in X,$$
(iii) for each $x \in X$, there is an element called the inverse of x such that: $x^{-1}x = xx^{-1} = e$.

(1.3)

The group operation is referred to as multiplication. A commutative group or Abelian group is one, where $xy = yx$ for all $x, y \in X$. The operation can then be written, for instance, as $(x\,y) \mapsto x + y$ (the inverse is then expressed by $-x$ and the identity denoted by 0). A subset A of X is a subgroup of X, if $x\,y \in A$ and $x^{-1} \in A$ for each $x \in A$, $y \in A$ or equivalently $xy^{-1} \in A$.

(iv) *Systems of sets, σ-algebra*

A system of sets S of a space X is referred to as a semi-ring, if, together with arbitrary sets A and A_1 in S, it also contains their intersections. Furthermore, if $A_i \subseteq A$, then the set
$$A = \bigcup_{i=1}^{n} A_i$$
of S, i.e. a finite number of disjoint sets $A_1, A_2, \ldots, A_n \in S$. The space X itself can also be written as the union of a countable number of disjoint sets $A_1, A_2, \ldots$ of S, i.e.:
$$X = \bigcup_i A_i.$$

The semi-ring of sets S is called a ring R, if together with arbitrary sets A_1, A_2 in S, it contains their sum $A_1 \cup A_2$. Thus, if S is an arbitrary semi-ring, the system of all sets $A \subseteq X$ that can be represented by
$$A = \bigcup_{i=1}^{n} A_i$$
of S is a ring. If the ring contains the entire space X, it is called an algebra. Thus an algebra $\mathscr{A}$ of sets includes with any set A its complement A' and with any sets $A_1, A_2, \ldots, A_n$ in $\mathscr{A}$, it contains the union
$$A = \bigcup_{i=1}^{n} A_i.$$

The algebra is a σ-algebra, if with every countable number of sets in $\mathscr{A}$ it also contains their union, i.e.:
$$A = \bigcup_{i=1}^{\infty} A_i.$$

The intersection of an arbitrary number of σ-algebras gives again a σ-algebra of sets in the space X. For any system of sets S, there exists a σ-algebra $\mathscr{A}$, which contains the system S. The smallest σ-algebra $\mathscr{A}$ containing the system of sets S is called the σ-algebra generated by S.

1.2.2 Topological and measurable spaces

(i) *Topological spaces*

A topological space is a set with a structure that permits the definition of the neighbourhood of points and the continuity of functions. A system S of subsets of a set X defines a topology on X, if the system S contains:

(i) the null set and the set X itself,
(ii) the union of each of its subsystems,
(iii) the intersection of everyone of its finite subsystems.

The sets in S are open sets of the topological space X. If X is a non-empty set and S consists of ϕ and X, the topology is called trivial. If X is any non-empty set and the open sets consist of all subsets of (ϕ and X included) the topology is called discrete.

A subset $A \subseteq X$ is open, iff it is contained in the topology of X. A set is closed, iff its complement is open. It is important for later considerations to consider the neighbourhood of a point $x \in X$ defined by a set $N(x)$ containing an open set that includes x. It is equally important to consider the neighbourhood of a subset $A \subset X$, which is the neighbourhood of every point of A. Thus a point $x \in X$ is called a limit point of $A \subseteq X$, if every neighbourhood $N(x)$ of x contains at least one point $a \in A$ different from x: $(N(x) - (x)) \cap A \neq \phi$ for all $N(x)$.

The notation supp f means the smallest closed set outside which the function f vanishes identically. It is to be noted that supp f is not the set $\{x : f(x) \neq 0\}$, but is the closure of this set. The closure of a set A or $\bar{A}$ is the intersection of all sets closed relative to the topology S on X. The interior of a set A is the largest open set contained in A. The set A is said to be dense in X, if $\bar{A} \supset X$. It is nowhere dense, if $\bar{A}$ has an empty interior. A space X is a topological space, if the system S of open sets is distinct with the previously given properties. In this context, the system S of X is a basis of X, if every open set is the union of open sets of S.

Thus a measurable topological space designated by $[X, \mathscr{A}]$ is one, in which the distinct σ-algebra $\mathscr{A}$ is generated by some basis of open sets of X.

The minimal σ-algebra that contains all open sets is called a Borel σ-algebra $\mathscr{A}$ of the space X. The sets $A \in \mathscr{A}$ are called Borel sets. This type of sets and their σ-algebra will be used extensively in the formulation of probabilistic mechanics. All countable unions and intersections of open and closed sets are

Borel sets. There exist, however, other kinds of sets and for their corresponding definitions see, for instance, Yosida [7] and others.

Amongst the basic topological spaces one can distinguish between a Hausdorff space (separated), a regular and a normal space. A topological space is Hausdorff, if any two distinct points possess disjoint neighbourhoods. A topological space is regular, if for any closed set A and any arbitrary point $x \notin A$ there exist disjoint open sets X_1 and X_2 such that $A \subseteq X_1$ and $x \in X_2$.

A topological space is normal, if any pair of disjoint closed sets have disjoint open neighbourhoods.

A topological space X is separable, if there exists some countable basis of open sets which is dense in X.

It will be convenient in the subsequent analysis of probabilistic mechanics to use subspaces of a more general topological space as well as product spaces.

(ii) *Product of spaces, compactness, connectedness, continuity*

The product of the spaces X_1 and X_2 is defined as the space, whose points are all possible pairs (x_1, x_2) in which $x_1 \in X_1$ and $x_2 \in X_2$. It is written as shown previously as $X = X_1 \times X_2$. A set A written as $A_1 \times A_2$ is called rectangular, i.e. it consists of all points $x = (x_1, x_2)$, where $x_1 \in A_1$ and $x_2 \in A_2$.

For two systems of sets S_1, S_2 that belong to the spaces X_1, X_2, respectively and where each of them is a semi-ring, the collection of all rectangular sets of the form $A = A_1 \times A_2$, where $A_1 \in S_1, A_2 \in S_2$ is also a semi-ring. Analogously, the product of two measurable spaces $[X_1, \mathscr{A}_1]$ and $[X_2, \mathscr{A}_2]$ is a measurable space $[X, \mathscr{A}]$; $X = X_1 \times X_2$ in which the σ-algebra $\mathscr{A}$ is the product of the σ-algebras $\mathscr{A}_1$ and $\mathscr{A}_2$. This means that $\mathscr{A}$ is generated by the semi-ring $\mathscr{A}_1 \times \mathscr{A}_2$ of all rectangular sets of the form $A = A_1 \times A_2$ with $A_1 \in \mathscr{A}_1$ and $A_2 \in \mathscr{A}_2$.

The product of two topological spaces X_1 and X_2 is the topological space $X = X_1 \times X_2$ in the sense of the above definitions, where $A = A_1 \times A_2$ with $A_1 \in S_1$, $A_2 \in S_2$ and where S_1, S_2 are bases of the open sets in X_1, X_2, respectively. In many instances use can be made of open sets for a given topology to construct simpler systems of sets on the basis of the concept of compactness. This involves, however, the notion of covering of the space.

Thus a system of open subsets $\{U_i\}$ of the space X is a covering, if each element in X belongs to at least one U_i, i.e.:

$$\bigcup_i U_i = X.$$

The covering $\mathscr{U}$ is locally finite, if for every point x there exists a neighbour-

1.2 Basic concepts and definitions of functional analysis 7

hood $N(x)$, that has a non-empty intersection with only a finite number of elements of $\mathcal{U}$.

A subset $A \subseteq X$ is compact, if it is Hausdorff and if every covering of A has a finite subcovering. Thus the subcovering of $\mathcal{U}$ is a subset of $\mathcal{U}$, which itself is a covering. The subset $A \subseteq X$ is said to be relatively compact, if the closure $\bar{A}$ is compact. A locally compact space has at every point a compact neighbourhood. Another concept is that of connectedness. A topological space is connected, iff the only subsets which are both open and closed are the null set and X itself.

Although certain basic definitions concerning mappings have already been given (paragraph 1.2.1(ii)) others pertaining to topological spaces are briefly mentioned here. Thus a function $f = f(x)$ on the topological space X with values in another space Y is a Borel function, if the inverse image $f^{-1}(B)$ of any open (or closed) set $B \subseteq Y$ is a Borel set of the space X.

The function $f = f(x)$ on X is continuous, if the inverse image $f^{-1}(B)$ of any open (or closed) set $B \subseteq Y$ is open or closed, respectively in X. The term homeomorphism means a bijection f, which is bi-continuous, i.e. f and f^{-1} are continuous. In a homeomorphism the images and inverse images of open sets are themselves open. The existence of a homeomorphism between topological spaces is an equivalence relation. A topological invariant is a property of a topological space, which is preserved under a homeomorphism. For example, the separation property, connectedness, compactness, etc. For a more detailed discussion on these concepts see, for instance, references [3, 4, 5].

(iii) *Metric spaces*

A metric space $[X, d]$ is a non-empty set of elements with a non-negative distance $d(x, y)$ defined for all pairs of elements or points $x, y \in X$ satisfying the following conditions for all points $x, y, z \in X$:

(i) $d(x, y) \geq 0$; $d(x, y) = 0$ iff $x = y$,
(ii) $d(x, y) = d(y, x)$ (symmetry), \qquad (1.4)
(iii) $d(x, y) \leq d(x, z) + d(y, z)$ (triangle inequality).

The distance function $d(x, y)$ is also called a metric. A neighbourhood $N_\delta(x)$ of a point x of the metric space X (δ being a positive number) is the set of all points y whose distance from x is smaller than δ. It is seen that the metric space is Hausdorff.

A subset A of the metric space X is open if, given a point $x \in A$, it contains some neighbourhood of this point. A function $f(x)$ on X is continuous, iff for every point $x \in X$ and a given positive number ε, a neighbourhood $N_\delta(x)$ of x is such that:
$$|f(x) - f(y)| \leq \varepsilon \quad \text{for all } y \in N_\delta(x). \qquad (1.5)$$
Considering now a closed set C in terms of the distance function from a point x

to C where
$$d(x, C) = \inf_{y \in C} d(x, y) \tag{1.6}$$

then the function $f(x) = d(x, C)$ is continuous on the space X and $f(x) = 0$ for $x \in C$ and $f(x) > 0$ for $x \notin C$. Hence in accordance with an earlier definition (paragraph 1.2.1(i)) a point x of an arbitrary set A of a metric space is a boundary point, if the distance from x to A and its complement is zero or if $d(x, A) = d(x, X \setminus A) = 0$. A metric space $[X, d]$ with a countable set of points $A = \{x_i; i = 1, 2, \ldots\}$, such that the closure $\bar{A} \supset X$ is called separable.

A sequence of points $x_1, x_2, \ldots$ of a topological space is convergent to a point x, if for every open set $0 \ni x$, there exists an "n" such that $x_m \in 0$ for $m > n$. A Cauchy sequence in the metric space $[X, d]$ of points $x_i; i = 1, 2, \ldots$ is such that $\lim d(x_i, x_j) = 0, i, j \to \infty$. The metric space is called complete, if each Cauchy sequence $\{x_i\}$ has a limit point, i.e.:

a point $x \in X$ so that $\lim_{i \to \infty} d(x_i, x) = 0$.

An arbitrary set A in a topological space X is said to be compact, if every open covering of A contains a finite subcover. If a separable topological space X is compact the distance $d(x_1, x_2)$ can be defined such that the neighbourhoods $N_\delta(x)$ form a basis of this space. A set A in X is compact, iff every sequence of points $x_i, i = 1, 2, \ldots$ of A contains a convergent subsequence. Thus, if X is an arbitrary compact space the system of all open Baire sets forms a basis in X (see Kuratowski [6] and Yosida [7]).

It is to be noted that by Urysohn's lemma [7] for any given disjoint closed sets A_1, A_2 of a compact space X, there exists a continuous real function $f(x)$ defined on this space with $0 \leqslant f(x) \leqslant 1$ for all $x \in X$ so that $f(x) = 0$ on A_1 and $f(x) = 1$ on A_2.

Another theorem due to Tychonoff [7] is of interest here. It concerns the product topology that will be used often in the following analysis. Thus, the product space of compact Hausdorff spaces is itself a compact space. A topological vector space is said to be metrizable, if there exists a metric d on X that induces the topology on X. One can distinguish two kinds of completeness for a topological vector space, i.e. the sequential completeness and completeness of the convergence of Cauchy nets [8]. In a metrizable topological vector space completeness, sequential completeness and metric completeness are equivalent (see, for instance, Kelley [8], [9]). A topological vector space that is metrizable and complete is called a Fréchet space. This type of space will be further discussed in the subsequent analysis.

(iv) *Linear spaces, semi-norms and locally convex linear spaces*

A space X is called linear, if for the elements $x \in X$ the operations of addition and multiplication by real numbers are such that

1.2 Basic concepts and definitions of functional analysis

(i) for any arbitrary numbers α, β and $x_1, x_2 \in X$, $x = \alpha x_1 + \beta x_2 \in X$
(ii) there is an element 0: $x + 0 = x$ and $\alpha \cdot 0 = 0$;
(iii) for any $x \in X$ there exists an inverse element denoted by $-x$ so that: $x + (-x) = 0$;
(iv) the following operations hold:

$$x_1 + x_2 = x_2 + x_1; \ 1 \cdot x = x; \ \alpha(\beta x) = (\alpha\beta)x$$
$$x_1 + (x_2 + x_3) = (x_1 + x_2) + x_3; \ (\alpha + \beta)x = \alpha x + \beta x;$$
$$\alpha(x_1 + x_2) = \alpha x_1 + \alpha x_2.$$

If α is a complex number X is called a complex linear space. The semi-norm of a vector in a linear space can be introduced on the basis of the theorem of locally convex spaces. Such spaces are defined by a system of semi-norms satisfying the separation axioms [3,4]. If this system reduces to a single semi-norm the corresponding linear space is a normed linear space. Since semi-norms are fundamental in the analysis concerned with linear topological spaces, the following definition is given:

Def. 1: A real valued function $f(x)$ defined on a linear space X is called a semi-norm on X, if the following conditions are satisfied:

(i) $f(x) = 0$, if $x = 0$,
(ii) $f(\alpha x) = |\alpha| f(x)$,
(iii) $f(x + y) \leqslant f(x) + f(y)$ (subadditivity). (1.7)

In addition a semi-norm also satisfies the following characteristic:

(iv) $f(x_1 - x_2) \geqslant |f(x_1) - f(x_2)|$ (1.8)

and in particular: $f(x) \geqslant 0$.

With reference to the above statements the following propositions can be given (for proof see Yosida [7]):

If $f(x)$ is a semi-norm on a topological space X and c an arbitrary positive number, then a set $M = \{x \in X; f(x) \leqslant c\}$ has the following properties:

(i) $M \ni 0$, M is convex: $x, y \in M$ and for $0 < \alpha < 1$,
(ii) $\alpha x + (1 - \alpha)y \in M$,
(iii) $x \in M$ and $|\alpha| \leqslant 1$ implies $\alpha x \in M$ (M is balanced),
(iv) for any $x \in M$, there exists $\alpha > 0$ so that $\alpha^{-1} x \in M$ (M is absorbing),
(v) $f(x) = \inf \alpha c$
$\quad \alpha > 0, \ \alpha^{-1} x \in M$. (1.9)

The functional $f_M(x) = \inf \alpha$
$\quad \alpha > 0, \ \alpha^{-1} x \in M$

is called the Minkowski functional of the set $M \subset X$. This functional $f_M(x)$ is therefore a semi-norm on X. In this context another definition is important, i.e. that of a locally convex linear space which is called a Fréchet space, if its topology is defined by just one semi-norm with the properties above. Hence, by using the above properties (1.9) the metric topology of such a space can be defined in terms of the distance $d(x, y)$ as follows:

Def. 2: $$d(x, y) = \|x - y\| \tag{1.10}$$

where this distance function satisfies the axioms of a metric given by (1.6).

It is shown, for instance, in reference [7], that for a bounded sequence

$$s\text{-}\lim_{n \to \infty} \alpha_n x_n = \alpha x, \text{ if } \lim_{n \to \infty} \alpha_n = \alpha \text{ and } s\text{-}\lim_{n \to \infty} x_n = x, \tag{1.11}$$

where s-lim designates the strong convergence limit. On this basis a linear space X can be referred to as a quasi-normed linear space, if for every $x \in X$, there is a number $\|x\|$ associated, called quasi-norm of the vector x such that:

$$\left.\begin{array}{l}
(1)\ \|x\| > 0 \text{ and } \|x\| = 0, \text{ iff } x = 0, \\
(2)\ \|x + y\| \leq \|x\| + \|y\|, \\
(3)\ \|-x\| = \|x\|,\ \lim_{\alpha_n \to 0} \|\alpha_n x_n\| = 0 \text{ and } \lim_{\|x_n\| \to 0} \|\alpha_n x_n\| = 0.
\end{array}\right\} \tag{1.12}$$

(v) *Banach spaces*

The linear space X is called a normed space, if for each of its elements x a norm $\|x\|$ is defined as a function of x. The mapping $x \mapsto \|x\|$ of a vector space X on $\mathbb{R}$ (field of real numbers) into $\mathbb{R}$ is a norm, if for $x \in X$ and any real number $\lambda \in \mathbb{R}$ the following properties hold:

(i) $\|x\| \geq 0$, $\|x\| = 0$, iff $x = 0$,
(ii) $\|\lambda x\| = |\lambda|\,\|x\|$,
(iii) $\|x + y\| \leq \|x\| + \|y\|$.

It should be noted that (i) differs from condition (i) defining a semi-norm. It is seen that the above norm induces a metric topology on X. However, not every metric space is necessarily a normed space. The vector space X together with a norm topology is called a normed topological vector space. A complete normed vector space is a Banach space or B-space. Obviously a Banach space is also a locally convex space.

1.2 Basic concepts and definitions of functional analysis

(vi) *Hilbert spaces*

The normed space is a Hilbert space, if a numerical function of two variables is defined such that the scalar product designated by $\langle x_1, x_2 \rangle$ satisfies the following conditions:

Def. 3: (i) $\langle x, x \rangle \geq 0$; $\langle x, x \rangle = 0$, iff $x = 0$,
(ii) $\langle x_1, x_2 \rangle = \overline{\langle x_2, x_1 \rangle}$
(iii) $\langle x, \alpha x_1 + \beta x_2 \rangle = \bar{\alpha} \langle x, x_1 \rangle + \bar{\beta} \langle x, x_2 \rangle$ (1.13)

for any α, β and elements $x_1, x_2 \in X$.

The norm $\|x\|$ of an element of the space X is defined by $\|x\| = \langle x, x \rangle^{1/2}$. A so-called pre-Hilbert space is a normed vector space with this norm and a complete pre-Hilbert space is a Hilbert space (see also reference [10]).

1.2.3 Measures, distributions in topological spaces

Before dealing with the main theorems of functional analysis and in particular with the important concepts of linear operators and semi-groups, which are required in the subsequent study, the fundamentals of measures in topological space will be considered first.

(i) *Measures*

A measure μ on a space X is a countably additive set function $\mu(A)$ on the σ-algebra $\mathscr{A}$ of the sets of the measurable space $[X, \mathscr{A}]$. This measure for any countable number of disjoint sets $A_1, A_2, \ldots \in \mathscr{A}$, $A = \bigcup_n A_n$ is defined by:

Def. 4: $$\mu(A) = \sum_n \mu(A_n). \tag{1.14}$$

The measure is finite, if $\mu(X) < \infty$ and σ-finite, if X can be represented by the union of countably many sets A_n: $\mu(A_n) < \infty$, $(n = 1, 2, \ldots)$. The triple $[X, \mathscr{A}, \mu]$ is then called a measure space. The probability space to be discussed subsequently is thus a measure space $[X, \mathscr{A}, \mu]$ with $\mu(X) = 1$.

A non-negative finite function $\mu = \mu(A)$ on a semi-ring S of sets in a space X is called a distribution, if for any set $A \in S$ which is

the union $\bigcup_n A_n$

the above given definition of $\mu(A)$ holds. If the semi-ring of sets S forms a ring, then the function μ is a distribution, iff it is finitely additive, i.e.:

$$\mu(A) = \sum_{i=1}^{n} \mu(A_i) \quad \text{for} \quad A = \bigcup_{i=1}^{n} A_i \tag{1.15}$$

with disjoint sets $A_1, A_2, \ldots, A_n \in S$ and continuous for a family $\{A_i\}$ such that $A_1 \supseteq A_2 \supseteq \ldots \supseteq A_n \supseteq \ldots$ and $\bigcap_n A_n = \phi$, (see [11]).

A weak distribution for $\mu(A)$ on S is one, where for any set $A \subseteq S$ as the union of a finite number of disjoint sets, the given definition of $\mu(A)$ holds. However, every weak distribution $\mu = \mu(A)$ on S can be extended to the ring of all sets $A \subseteq X$ that are unions of a countable number of disjoint sets of S. Such an extension is unique and is defined by

$$\mu(A) = \sum_{i=1}^{n} \mu(A_i).$$

Considering now a bounded measure $\mu(A)$ on the semi-ring of sets S in X, where $\mu(X) < \infty$, then the exterior measure of a set $A \subseteq X$ is defined by a number

$$\mu'(A) = \inf_i \sum \mu(A_i).$$

Here the infimum is taken over all countable sets $A_1, A_2, \ldots \in S$ covering A. On the other hand, if $A_1^*, A_2^*, \ldots \in S$ is a family of sets contained in A, then $\mu^*(A) = \sup \sum_i \mu(A_i^*)$ is the interior measure of this set. If $\mu' = \mu^*$ in a topological space it is called a regular measure. The number $\mu'(A)$ is finite for every set $A \subseteq X$. The set A is said to be measurable with respect to the measure μ if $\mu'(A) = \mu'(X) - \mu'(X \setminus A)$. It is to be noted that for a given set there exists a number m between the bounds $\mu'(A) > m \geqslant \mu'(X) - \mu'(X \setminus A)$ and an extension μ_1 of μ such that the measure $\mu_1(A) = m$. Equivalent definitions are given in other texts (see, for instance, [5,11]). Of further interest is the so-called product measure to be employed in the later analysis.

Thus for two σ-finite measurable spaces $[X, \mathcal{A}, \mu]$ and $[X^*, \mathcal{A}^*, \mu^*]$ one can define the σ-field $\mathcal{A} \otimes \mathcal{A}^*$ of subsets of the direct product $X \otimes X^*$ as the smallest σ-field containing all sets of the form $A \times A^*$, $A \in \mathcal{A}$, $A^* \in \mathcal{A}^*$. It can be proved, that there exists a unique measure $\mu \otimes \mu^*$ defined on $\mathcal{A} \otimes \mathcal{A}^*$ such that for all $A \in \mathcal{A}$ and $A^* \in \mathcal{A}^*$ one has

$$(\mu \otimes \mu^*)(A \times A^*) = \mu(A)\mu^*(A^*). \tag{1.16}$$

The corresponding measure space $[X \times X^*, \mathcal{A} \otimes \mathcal{A}^*, \mu \otimes \mu^*]$ is then called the product measure space of X and X^*. The term measurable almost everywhere a.e., is a characteristic to hold for all points $x \in X$ except perhaps for points of a set $A \subseteq X$ of measure $\mu(A) = 0$. Extending the class of measurable sets such that every subset of a set with measure zero is measurable, and has the measure zero yields a completed measure from the original one. Any measure $\mu(A)$ can be extended to a complete measure.

The positive measure on Borel sets of a locally compact topological space

1.2 Basic concepts and definitions of functional analysis 13

(Hausdorff) is called a Borel measure. Another measure defined on $\mathbb{R}$ is the Lebesgue measure. It is a regular Borel measure on $\mathbb{R}$, which is invariant by translation. For defining the regularity property consider X to be a topological space and the σ-algebra $\mathscr{A}$ of Baire sets. A finite measure $\mu(A)$ on $\mathscr{A}$ is then regular, if for every Baire set A: $\mu(A) = \inf \mu(U)$ where the inf is taken over all open Baire sets $U \supseteq A$ or equivalently $\mu(A) = \sup \mu(C)$ where the sup is taken over all compact sets $C \subseteq A$. If the space X is the real line, the Baire sets and Borel sets coincide. One can also consider the Lebesgue measure l on $\mathbb{R}^n$. It is then the product measure $l \otimes l \otimes \ldots \otimes l$ and the only regular Borel measure that is invariant by translation and rotation. For a more detailed discussion on the existence and uniqueness of these measures see, for instance, Halmos [11].

(ii) *Measurable and integrable functions*

If $[X, \mathscr{A}, \mu]$ is an arbitrary space with a finite complete measure on the σ-algebra $\mathscr{A}$, then a function $f = f(x)$ on X with values in a space Y, is called simple, if it has a denumerable number of values $y_1, y_2, \ldots$ on the disjoint sets $A_1, A_2, \ldots \in \mathscr{A}$. A real function on $[X, \mathscr{A}, \mu]$ is measurable, if it is a measurable map from X into $\mathbb{R}$, i.e. if $\{x; a < f(x) < b\} \in \mathscr{A}$ for $\forall a, b \in \mathbb{R}$. If f and g are measurable functions on X and an arbitrary number $\lambda \in \mathbb{K}$, then the functions $\lambda f, f + g, fg, |f|$ are also measurable. These properties, however, are not always satisfied for mappings into arbitrary spaces. They do hold for mappings of a σ-finite measure space $[X, \mathscr{A}, \mu]$ into a separable Banach space (see also ref. [9]).

A real valued simple function $f(x)$ on X is integrable with respect to the measure μ, if the series $\sum_i |y_i| \mu(A_i)$ converges. One can then define the integral over X by:

$$\int_X f(x) \mu(dx) = \sum_i y_i \mu(A_i). \tag{1.17}$$

An arbitrary real function $f(x)$ on X is integrable, if it is the limit of a uniformly converging sequence of simple integrable functions, i.e.:

$$f(x) = \lim_{n \to \infty} f_n(x); \quad \int_X f(x) \mu(dx) = \lim_{n \to \infty} \int_X f_n(x) \mu(dx). \tag{1.18}$$

The sequence of measurable functions on $[X, \mathscr{A}, \mu]$, which are finite a.e., is said to converge in measure or $f_n(x) \overset{\mu}{\rightsquigarrow} f(x)$ to the measurable function $f(x)$, if

$$\lim \mu(\{x: |f_n(x) - f(x)| > \varepsilon\}) = 0 \quad \text{for every} \quad \varepsilon > 0. \tag{1.19}$$

14 Mathematical Preliminaries

Hence, from the above statements it follows that a function $f(x)$ is integrable, iff it is measurable and its absolute value $|f(x)|$ is integrable. It is to be noted that by Egorov's theorem [4], if the sequence $f_n(x)$ converges point-wise a.e. in a finite measure μ on X to $f(x)$ then for every $\varepsilon > 0$ there exists a subset A of $X : \mu(X \setminus A) \leqslant \varepsilon$ and the convergence f_n to f is uniform on A (see also Yosida [7]). There are of course other measures in topological spaces as, for instance, the generalized measure of a set A, that involves the concept of the variance as a measure on $\mathscr{A}$. Probability measures will be discussed in later sections and for a more comprehensive study of measures reference is made to various texts listed in the bibliography of this volume [5, 11, 12].

1.2.4 Linear operators and semi-groups

As mentioned earlier the formulation of the probabilistic mechanics frequently uses the concepts of linear bounded operators and the corresponding semi-groups. It is therefore necessary to discuss briefly the fundamental notions of linear functionals, semi-norms and semi-groups.

(i) Linear functionals

Considering a linear space U and a real (or complex) valued function $f : U \to \mathbb{R}^1$ on this space, then f is called a linear functional, if $f(\alpha u_1 + \beta u_2) = \alpha f(u_1) + \beta f(u_2)$ for all real (or complex) values α, β and $u_1, u_2 \in U$.

It is usually assumed, if U is a linear topological space, that f is continuous. If U is a complete Hilbert space, then every linear continuous functional defined on U has the form:

$$f(u) = \langle u, x \rangle; \quad u \in U \qquad (1.20)$$

where x is a fixed element of U and $\langle \cdot, \cdot \rangle$ denotes here the scalar product. On an arbitrary normed space U a linear functional is continuous, iff it is bounded, i.e.:

$$\|x\| = \sup_{\|u\|=1} |\langle u, x \rangle| < \infty \qquad (1.21)$$

where the supremum is taken for all elements $u \in U$ with $\|u\| = 1$. The number $\|x\|$ is the norm of the linear functional $f = \langle u, x \rangle$. For a specific u_0 of the normed space U, there exists an f such that $\langle u, x \rangle$, $u \in U$ and $\langle u_0, x \rangle$, $u_0 \in U_0$, U_0 being a subspace of U. Hence, one can extend this definition to a continuous linear functional f on U by:

$$\|x\| = \sup_{\|u\|=1, u \in U_0} |\langle u, x \rangle| \qquad (1.22)$$

in which the supremum is taken only for the elements u from $U_0 \subseteq U$, $\|u\| = 1$.

1.2 Basic concepts and definitions of functional analysis

(ii) *Linear operators*

Considering a linear space U and a function $v = Tu$ on this space ($u \in U$) with values in another linear space V, then the linear operator T on U is defined for the relation $v = Tu$, if

$$T(\alpha u_1 + \beta u_2) = \alpha T u_1 + \beta T u_2 \qquad (1.23)$$

for any $u_1, u_2 \in U$ and any numbers α, β. The operator T^{-1} defined on the elements of the space V is referred to as the inverse operator, if $T^{-1}v = u$ for $v = Tu$. It is common usage to denote by $\mathscr{D} = \mathscr{D}(T)$ the domain of the operator, by $R(T) = \{v \in V; v = Tu, u \in \mathscr{D}(T)\}$ the range of the operator and by $N(T) = \{u \in \mathscr{D}(T); Tu = 0\}$ the null space. T is also called a linear transformation (mapping) on $\mathscr{D}(T) \subseteq U$ into V. If the range $R(T)$ of the operator is contained in the scalar field $\mathbb{K}$ for example, then T becomes a linear functional on $\mathscr{D}(T)$. If the operator T gives a one-to-one map of $\mathscr{D}(T)$ on $R(T)$, then the inverse map T^{-1} gives a linear operator on $R(T)$ onto $\mathscr{D}(T)$. For two linear operators with domains $\mathscr{D}(T_1)$, $\mathscr{D}(T_2)$ contained in a linear space U and having the ranges $R(T_1), R(T_2)$ in V, respectively, the operators will be equal, iff $\mathscr{D}(T_1) = \mathscr{D}(T_2)$ and $T_1 u = T_2 u$ for all $u \in \mathscr{D}(T_1) = \mathscr{D}(T_2)$. If $\mathscr{D}(T_1) \subseteq \mathscr{D}(T_2)$ and $T_1 u = T_2 u$ for all $u \in \mathscr{D}(T_1)$, the operator T_2 is called an extension of T_1 and T_1 a restriction of T_2. The operator T on a linear topological space U mapping $U \to V$ is continuous, if the function $v = Tu$ is continuous. If the spaces U, V are normed spaces, then T is linear and continuous, iff it is linear and bounded, i.e.:

$$\|T\| = \sup_{\|u\|=1} \|Tu\| < \infty \qquad (1.24)$$

where the supremum is taken for all elements $u \in U$ with $\|u\| = 1$ and $\|T\|$ is the norm of T. If T is a linear operator on a Banach space X into itself such that $x, y \in X$ and α, β are real (or complex) scalars, then: $T(\alpha x + \beta y) = \alpha T x + \beta T y$ and the operator will be bounded similar to the case of a linear functional, if

$$\|T\| = \sup_{x \neq 0, x \in X} \frac{\|Tx\|}{\|x\|} < \infty. \qquad (1.25)$$

In certain applications the term projection is used. Thus a bounded linear operator S on the Banach space X is called a projection, if $S^2 = S$. The operator S determines the subspaces X_1, X_2 of the Banach space X with $X_1 = \{x \in X; Sx = x\}$ and $X_2 = \{x \in X; Sx = 0\}$. For a given $x \in X$ one can write that $x = Sx + (x - Sx)$ with $Sx \in X_1$ and $(x - Sx) \in X_2$. It is usual to call S a projection of X on X_1. It is of interest for later work to refer to a linear operator T as a contraction operator, if $\|T\| \leq 1$. For two given bounded linear operators T_1 and T_2 on Banach spaces X_1 and X_2 with $X_1 \subset X_2$, the operator T_2 is referred to as a dilatation of T_1, if there is a projection operator S that projects X_2 on X_1 so that $T_1 = ST_2 S$.

If T is a bounded linear operator on a Hilbert space H, there is a bounded linear operator T^* on H such that:

$$(T^*x, y) = (x, Ty) \text{ for all } x, y \in H. \tag{1.26}$$

The operator T^* is called the adjoint of T and $\|T\| = \|T^*\|$. If $T = T^*$, the operator T is called a self-adjoint operator, which is frequently used in mechanics. Further considerations to bounded linear operators will be given in later sections after the notions of probability and their measures on Borel fields are introduced.

(iii) *Continuity and boundedness of operator*

Consider two linear spaces X and Y over the same scalar field ℼ and a linear operator T on $\mathscr{D}(T) \subseteq X$ into Y, it will be continuous everywhere on $\mathscr{D}(T)$, iff it is continuous at the zero vector $x = 0$. As a consequence of the given definition of semi-norms the following theorem holds:

"If X, Y are two locally convex spaces with semi-norms $\{p\}$, $\{p'\}$, respectively, that define the topologies of X, Y a linear operator T on $\mathscr{D}(T) \subseteq X$ into Y is continuous, iff for every semi-norm $p \in \{p\}$ there exists a semi-norm $p' \in \{p'\}$ and a positive number α'' such that:

$$p'(Tx) \leqslant \alpha p(x) \quad \text{for} \quad \forall x \in \mathscr{D}(T). \tag{1.27}$$

If T is a continuous linear operator on a normed linear space X into a normed linear space Y one defines:

$$\|T\| = \inf_{\alpha \in X} \alpha; \, X = \{\alpha; \|Tx\| \leqslant \alpha \|x\| \quad \text{for} \quad \forall x \in X\}. \tag{1.28}$$

Invoking (1.27) or rather its corollary, i.e. that an inverse of the operator T exists, iff there is a positive constant β such that:

$$\|T\| \geqslant \beta \|x\| \quad \text{for} \quad \forall x \in \mathscr{D}(T) \tag{1.29}$$

it is readily seen that:

$$\|T\| = \sup_{\|x\|=1} \|Tx\| \tag{1.30}$$

in which $\|T\|$ is the norm of the operator. Thus, a continuous linear operator on a normed linear space X into Y is a bounded linear operator on X into Y. Linear operators on finite dimensional normed vector spaces are necessarily continuous. However, this is not the case for infinite dimensional spaces and hence considerations of the theory of semi-group of operators becomes significant in this context.

1.2 Basic concepts and definitions of functional analysis 17

(iv) *Semi-groups of linear operators*

For the purpose of later considerations, that are concerned with the introduction of probabilistic concepts to the mechanics of discrete media and in particular the modelling of the latter on the basis of the Markov Theory, a brief discussion at this point of the theory of semi-groups of bounded linear operators and its application to stochastic processes may be indicated. Thus certain theorems and definitions which are due mainly to the work of E. Hille [13] and K. Yosida [7] are given below. The proposition by Hille concerning continuous linear operators on locally convex linear topological spaces may be stated as follows:

"If X is a Banach space and T_t a general linear bounded operator function for $t \geq 0$, the one-parameter family of bounded linear operators $L(X, X)$ satisfies the semi-group property", i.e.:

$$\left.\begin{array}{l}\text{(i) } T_{t+s} = T_t T_s \text{ for } t, s > 0, \\ \text{(ii) } T_0 = I \quad \text{(Identity operator).}\end{array}\right\} \quad (1.31)$$

By a one-parameter linear semi-group in a real or complex Banach space X is meant the family $\{T_t; t \geq 0, t \in \mathbb{R}^1 \text{ (real line)}\}$ of operators in X satisfying the conditions in (1.31). Evidently, if the operator T_t is an automorphism one has $T_{-t} = T_t^{-1}$ and the family is a group. If every $T_t (t \geq 0)$ is a bounded operator in X and satisfies the conditions (1.31), one can define a strong limit in the Banach space X in the form of:

$$s\text{-}\lim_{t \to t_0} T_t x = T_{t_0} x \quad \text{for all} \quad t_0 \geq 0 \quad \text{and all} \quad x \in X. \quad (1.32)$$

In this case $\{T_t\}$ is called a semi-group of the C_0-class (Yosida [7]). The one-parameter linear semi-group $\{T_t; t \geq 0\}$ is contracting, if for all $t \geq 0$, one has:

$$\|T_t x\| \leq \|x\|, \quad x \in X \quad (1.33)$$

In subsequent considerations by employing one-parameter linear semi-groups in the Banach space X, the notion of an infinitesimal generator of the semi-group is important. It designates by definition an operator $L: X \to X$ of the form:

$$Lx = s\text{-}\lim_{t \downarrow 0} t^{-1} (T_t - I) x. \quad (1.34)$$

In view of Hille's proposition (see also Yosida [7]) the C_0-class of semi-groups $\{T_t\}$ will satisfy the condition that:

$$\|T_t\| \leq M e^{\beta t} (0 \leq t < \infty)$$

and where $M > 0$ and $\beta < \infty$ are constants. Multiplying above by $e^{-\beta t}$ shows that one may assume the semi-group $\{T_t\}$ of the C_0-class to be equibounded, i.e.:

$$\|T_t\| \leq M \quad \text{for} \quad 0 \leq t < \infty. \quad (1.35)$$

In particular, if $M < 1$ and if $\|T_t\| \leq 1$ $(0 \leq t < \infty)$, $\{T_t\}$ is said to be a contraction group of the C_0-class. The strong continuity condition of (1.32) can be stated in an equivalent form or weaker form for the family $\{T_t; t \geq 0\}$ satisfying (1.31), that is closely related to the theory of Markov processes as follows:

$$w\text{-}\lim_{t \downarrow 0} T_t x = x, \quad \text{for} \quad \forall x \in X \tag{1.36}$$

or the weak limit of the bounded operator T_t (for the proof see, for instance, Yosida [7]). A more general class of semi-groups on a locally convex linear topological space X is obtained, if $\{T_t; t \geq 0\}$ satisfies the conditions:

(i) $T_{t+s} = T_t T_s$; $T_0 = I$,

(ii) $\lim_{t \to t_0} T_t x = T_{t_0} x$ for any $t_0 \geq 0$, $x \in X$ $\qquad$ (1.37)

and the family of mappings $\{T_t\}$ is equi-continuous in time, i.e. if for any continuous semi-norm p on X, there exists a continuous semi-norm p' on X such that $p(T_t x) \leq p'(x)$ for all $t \geq 0$ and all $x \in X$. Every C_0-semi-group $\{T_t\}$ on X has an infinitesimal generator A, which is defined as:

$$Ax = \lim_{h \downarrow 0} h^{-1}(T_h x - x) \quad \text{or} \quad \lim_{h \downarrow 0} h^{-1}(T_h - I)x. \tag{1.38}$$

This definition arises from the assumption of an equi-continuous semi-group of class C_0 defined on a locally convex linear topological space X, to be sequentially complete. The operator A is linear with a domain

$$\mathscr{D}(A) = \{x \in X; \lim_{h \downarrow 0} h^{-1}(T_h - I)x \text{ exists in } X\},$$

where $\mathscr{D}(A)$ is a non-empty set. It contains at least the vector zero. This domain, however, is larger and can be shown to be dense in X. Conversely, if L is any linear operator on a dense subspace of X, then an abstract Cauchy problem for L in classical mechanics consists of finding a C_0-semi-group, whose infinitesimal generator is an extension of L. Thus, for instance, if L is a linear differential operator, a C_0-semi-group on X, whose infinitesimal operator is an extension of L is called a solution semi-group in X of the abstract Cauchy problem, that is given by a system of partial differential equations representing evolution equations in applications of classical mechanics.

1.3 PROBABILITY, RANDOM VARIABLES AND PROCESSES

In order to establish the mathematical foundations of probabilistic mechanics, certain concepts and definitions in addition to those given in the preceding sections are required. Although several theories have been proposed

1.3 Probability, random variables and processes

in the past on the basis of statistical mechanics and the notions of functional analysis, no attempt has been made to introduce probability and stochastic processes in a physical theory concerned with the representation of the behaviour of discrete materials. In general, one can distinguish two kinds of physical quantities:

(i) Those whose results are usually thought to be real numbers or real and or complex functions, apart from certain errors due to the experimental procedures involved in their determination.
(ii) Those whose results are probability distributions of such numbers and functions involving the precision with which they can be established.

The quantities of the first kind are conventionally termed deterministic, whilst those of the second type are called probabilistic. It is the latter type of quantities that are considered to be fundamental in mechanics and hence the deterministic ones are regarded only as a special case of the probabilistic ones.

1.3.1 Probability

(i) *Definitions of probability; probability space and measure*

Among several intuitive ways of defining probability, there are two distinct definitions. One is based on the frequency interpretation due to von Mises [14] and the other on the axioms of Kolmogorov [1], which makes probability theory part of measure theory.

A more recent discussion on the foundation of probability and the notions of absolute probability and frequency has been given by van Fraassen [15]. The basic definitions concerning classical probability and quantum probabilities are treated amongst others by Gudder [16].

In probability theory in general a random experiment is called a trial. If such an experiment is carried out without any prior knowledge the result is an outcome ω. The set of all possible outcomes $\{\omega\}$ is the sample space Ω.

An event E is a subset of Ω such that $E = \{\omega \in \Omega; \omega$ satisfying certain properties$\}$. Hence the statement that an event E occurs means that the outcome of the random experiment is an element of E or $\omega \in E$. Events may be simple (indecomposable) or compound and then they are decomposable. A simple event has always only one outcome. The empty or null set ϕ contains no outcome and hence is an impossible event. It is postulated, that an event is one for which a probability of occurrence can be specified. In terms of the above-mentioned frequency interpretation, if an experiment is carried out n-times under identical conditions the probability of the event E_i is given by:

$$\mathscr{P}\{E_i\} = \lim_{n \to \infty} \frac{n_i}{n} \qquad (1.39)$$

where n_i is the occurrence of the events E_i. However, whilst n may be large it never becomes infinite. Thus, the definition adopted here is that based on Kolmogorov's axioms in probability theory [1], which postulates that the relation (1.39) holds almost everywhere, i.e.:

$$\mathscr{P}\{E_i\} \stackrel{a.e.}{=} \lim_{n \to \infty} \frac{n_i}{n}. \tag{1.40}$$

This definition is based on the following axioms:

A.1: $\mathscr{P}\{E_i\} \geq 0$ (0 being an impossible event),
A.2: $\mathscr{P}\{\Omega\} = 1$ (1 being always a certain event),
A.3: For two mutually exclusive events E_i, E_j
$\mathscr{P}\{E_i \cup E_j\} = \mathscr{P}\{E_i\} + \mathscr{P}\{E_j\}; E_i \cap E_j = \phi, i \neq j.$
$\tag{1.41}$

The notation $\mathscr{P}\{E\}$ or the probability of occurrence of the event refers to the probability measure of E or simply the probability of E. As mentioned in section 1.2.1 a non-empty class of sets satisfying the conditions of a field can be given to the events E_i, i.e.

$$\mathscr{F} = \{E_i; i = 1, 2, \ldots\}$$

such that $E_i \in \mathscr{F} \Rightarrow \bigcup_{i=1}^{\infty} E_i \in \mathscr{F}$. More specifically a σ-field is a non-empty class of event sets E_i, where:

(i) $E_i \in \mathscr{F} \Rightarrow E_i' \in \mathscr{F}; i = 1, 2, \ldots,$
(ii) $E_i \in \mathscr{F}$ and $E_j \in \mathscr{F} \Rightarrow E_i \cap E_j \in \mathscr{F},$
(iii) $E_i \in \mathscr{F} \Rightarrow \bigcup_{i=1}^{\infty} E_i \in \mathscr{F}.$
$\tag{1.42}$

Thus, a σ-field $\mathscr{F}$ is closed under the operations of countable intersection, union and complement. Using the above postulates of a σ-field an axiomatic definition of probability can be given as follows:

I. $\mathscr{P}\{E_i\} \geq 0$, $\mathscr{P}\{E_i\} = 0$ if $E_i = \phi$,
II. $\mathscr{P}\{\Omega\} = 1$,
III. $\mathscr{P}\left\{\bigcup_{i=1}^{\infty} E_i\right\} = \sum_{i=1}^{\infty} \mathscr{P}\{E_i\}$ if $E_i \cap E_j = \phi, i \neq j.$
$\tag{1.43}$

It is important to note that a probability measure $\mathscr{P}$ is a *set function*, since its argument is, in general, a set that may contain more than one point. Furthermore, the probability associated with the sample space Ω as a whole, is unity. Considering the sample space Ω that contains all possible outcomes together with the above defined σ-field $\mathscr{F}$ of the events, it will be a measurable space $[\Omega, \mathscr{F}]$. Hence a probability space or equivalently a probabilistic measure space is a measurable space $[\Omega, \mathscr{F}]$ together with an appropriate

1.3 Probability, random variables and processes

probability measure $\mathscr{P}$ given by the axiomatic definition in (1.43). It is then represented by the triple $[\Omega, \mathscr{F}, \mathscr{P}]$, where $\mathscr{P}$ is a set function defined for events and not for outcomes. Although a more comprehensive study of probability measures will be given in later sections, it may be stated here that generally the measure $\mathscr{P}$ on the space Ω is a countably additive set function $\mathscr{P}\{E\}$ on the σ-algebra $\mathscr{F}$ of the sets of the measurable space $[\Omega, \mathscr{F}]$. This measure for any countable number of disjoint sets

$$E_1, E_2, \ldots \in \mathscr{F}, E = \bigcup_n E_n$$

is defined by:

$$\mathscr{P}\{E\} = \sum_n \mathscr{P}\{E_n\} \tag{1.44}$$

The measure is finite, if $\mathscr{P}\{\Omega\} < \infty$ and σ-finite, if Ω can be represented by the union of countably many sets $E_n: \mathscr{P}\{E_n\} < \infty$ $(n = 1, 2, \ldots)$. The triple $[\Omega, \mathscr{F}, \mathscr{P}]$ is then a measure space.

In this context the notion of a conditional probability space is considered by some authors as more fundamental in the mathematical theory of probability (see, for instance, Rényi [17], Kolmogorov [2] and others). This concept originated from the fact that the probability of an event depends largely on the circumstances under which the occurrence or non-occurrence of such an event is observed. As a consequence conditional probability spaces are introduced, which are generated by bounded measures so that the given definitions of probability spaces and measures should be regarded as a special case of a more general theory of probability. For the introduction of conditional probability arguments that will be used in later sections of this text the following definitions may be given.

(ii) *Conditional probability and independence of events*

Using the frequency interpretation of probability (1.39) and considering two events E_1 and E_2 first the concept of a joint probability will be given. If n_1 denotes the number of outcomes favourable to the event E_1 in an experiment and n_2 the number of outcomes favourable to the event E_2 then:

$$\mathscr{P}\{E_1\} = \frac{n_1}{n_\Omega}; \quad \mathscr{P}\{E_2\} = \frac{n_2}{n_\Omega} \tag{1.45}$$

where n_Ω is the number of total outcomes. Hence, if E_1 and E_2 are not mutually exclusive, $E_1 \cap E_2 \neq \phi$. In that case, if n_{12} is the number of outcomes that are favourable to E_1 and E_2 the joint probability is defined by:

$$\mathscr{P}\{E_1 \cap E_2\} = \frac{n_{12}}{n_\Omega} \tag{1.46}$$

22 Mathematical Preliminaries

$\mathscr{P}\{E_1\}$, $\mathscr{P}\{E_2\}$ are usually referred to as marginal probabilities of E_1 and E_2, respectively. The conditional probability of an event E_1 with respect to another E_2 denoted by $\mathscr{P}\{E_1|E_2\}$ is the probability that E_1 occurs under the condition that E_2 has already occurred. Thus, only the outcomes favourable to E_2 are considered and not all outcomes. In terms of the frequency definition of probability, it is then represented by:

Analogously:

$$\mathscr{P}\{E_1|E_2\} = \frac{n_{12}}{n_2} = \frac{n_{12}/n_\Omega}{n_2/n_\Omega} = \frac{\mathscr{P}\{E_1 \cap E_2\}}{\mathscr{P}\{E_2\}}. \tag{1.47}$$

$$\mathscr{P}\{E_2|E_1\} = \frac{n_{12}}{n_1} = \frac{n_{12}/n_\Omega}{n_1/n_\Omega} = \frac{\mathscr{P}\{E_1 \cap E_2\}}{\mathscr{P}\{E_1\}}. \tag{1.48}$$

It is readily seen that from these relations one also obtains:

$$\mathscr{P}\{E_1|E_2\} = \frac{\mathscr{P}\{E_2|E_1\}\mathscr{P}\{E_1\}}{\mathscr{P}\{E_2\}} \tag{1.49}$$

which is the well-known Bayes rule in probability. Considering n mutually exclusive sets E_i, $i = 1, 2, \ldots, n$ in Ω such that

$$\sum_{i=1}^{n} E_i = \Omega$$

and if E is an arbitrary event in Ω it follows that:

$$\mathscr{P}\{E\} = \sum_{i=1}^{n} \mathscr{P}\{E|E_i\}\mathscr{P}\{E_i\}. \tag{1.50}$$

Two events are said to be pair-wise independent, if

$$\mathscr{P}\{E_i \cap E_j\} = \mathscr{P}\{E_i\}\mathscr{P}\{E_j\}; \; i \neq j. \tag{1.51}$$

However, it does not follow strictly that pair-wise independence implies absolute independence. The latter is characterized by:

$$\mathscr{P}\{E_1 \cap E_2 \cap \ldots \cap E_n\} = \mathscr{P}\{E_1\}\mathscr{P}\{E_2\}\ldots\mathscr{P}\{E_n\} \tag{1.52}$$

It should be noted, that the above relation may be true without the events E_i being independent of each other as a system. Thus in order that all events E_i to be independent the following relations would have to be satisfied:

$$\mathscr{P}\{E_i \cap E_j\} = \mathscr{P}\{E_i\}\mathscr{P}\{E_j\},$$
$$\mathscr{P}\{E_i \cap E_j \cap E_k\} = \mathscr{P}\{E_i\}\mathscr{P}\{E_j\}\{E_k\},$$
$$\vdots \qquad \vdots \; \vdots$$
$$\mathscr{P}\{E_1 \cap E_2 \cap \ldots \cap E_n\} = \mathscr{P}\{E_1\}\mathscr{P}\{E_2\}\ldots\mathscr{P}\{E_n\},$$

1.3 Probability, random variables and processes

but for two events only equation (1.51) will hold and leads to:

$$\mathscr{P}\{E_1|E_2\} = \mathscr{P}\{E_1\}; \ \mathscr{P}\{E_2|E_1\} = \mathscr{P}\{E_2\}$$

from which it may be concluded that they are mutually independent.

1.3.2 Random variables and their distributions

(i) *Random variables*

As stated earlier the outcomes of physical phenomena, when considered as random, can be represented within the classical probability theory by random numbers. However, these outcomes more generally may require the representation in terms of random variables, whose values are in topological vector spaces. Following the modern trend of probabilistic functional analysis [18] one can define random variables or random functions in a general topological space. Since the theory of random variables in spaces other than Banach spaces is as yet not fully developed and most of the field quantities considered in probabilistic mechanics are sufficiently represented in this setting, the following definition of a random variable is given:

Def. 1: "A mapping $x: \Omega \to \mathscr{X}$ is called a random variable with values in $\mathscr{X}$, if the inverse image under x of every Borel set E belongs to $\mathscr{B}$; that is $x^{-1}(E) \in \mathscr{B}$ for all $E \in \mathscr{F}$, where $[\Omega, \mathscr{B}, \mathscr{P}]$ is the probability space and $[\mathscr{X}, \mathscr{F}]$ the Banach space with σ-algebra $\mathscr{F}$ of the Borel sets E of $\mathscr{X}$."

It is easily recognized that, if $\mathscr{X}$ is identical to $\mathbb{R}^1$ this definition reduces to the conventional definition of a real valued random variable. It is important to note that in probabilistic mechanics a random variable designated by $x = x(\omega)$ can be considered in most cases to be "directly given" [19], if each elementary outcome $\omega \in \Omega$ corresponds to a point $\xi \in \mathscr{X}$ in the probabilistic function space. Hence accordingly the sample space Ω will be identified with the probabilistic function space $\mathscr{X}$, i.e. $\Omega = \mathscr{X}$. It follows that the corresponding σ-algebras $\mathscr{B}$ and $\mathscr{F}$ are also the same together with the associated measures. In the following the probabilistic measure on the measurable topological space $[\mathscr{X}, \mathscr{F}]$ will be denoted by $\mathscr{P}$ and the above identification will be employed whenever possible. There are two basic types of real valued random variables, i.e. discrete and continuous random variables. A discrete variable $x = x(\omega)$ takes only a finite or denumerable number of different values with corresponding probabilities $P_x(\xi) = \mathscr{P}\{x = \xi\}$, where $\{x = \xi\}$ indicates, that the event for the random variable x takes the value ξ or $\{\omega: x(\omega) = \xi\}$. The probability of the event $\xi_1 \leqslant x \leqslant \xi_2$ that x takes one of the values lying between $\xi_1 \leqslant \xi \leqslant \xi_2$ is then

$$\mathscr{P}\{\xi_1 \leqslant x \leqslant \xi_2\} = \sum_{\xi_1}^{\xi_2} P_x(\xi).$$

24   Mathematical Preliminaries

The summation extends over the finitely countable number of values that the discrete random variable may take. The notation $P_x(\xi)$ reflects, that it is a function of all possible values ξ of x.

The function $F_x(\xi)$ of an arbitrary random variable x defined for all values of ξ on the real line $\mathbb{R}^1$ such that:

$$F_x(\xi) = \mathscr{P}\{x \leqslant \xi\}; \quad (-\infty < \xi < \infty)$$

is called the distribution function of x. For any value ξ_1 and ξ_2, $(\xi_1 < \xi_2)$ one has therefore:

$$\mathscr{P}\{\xi_1 \leqslant x \leqslant \xi_2\} = F_x(\xi_2) - F_x(\xi_1). \tag{1.53}$$

The distribution function $F_x(\xi)$ of a discrete random variable is piecewise constant. If the distribution function of x is absolutely continuous and differentiable, then

$$p_x(\xi) = \frac{\mathrm{d}}{\mathrm{d}\xi} F_x(\xi)$$

represents the density of the probability distribution or briefly the probability density of the continuous distribution of the random variable x. The probability density is a non-negative function $p_x(\xi)$ such that for any ξ_1, ξ_2 with $\xi_1 < \xi_2$:

$$\mathscr{P}\{\xi_1 \leqslant x \leqslant \xi_2\} = \int_{\xi_1}^{\xi_2} p_x(\xi)\mathrm{d}\xi \tag{1.54}$$

The functions $F_x(\xi)$, $P_x(\xi)$ and $p_x(\xi)$ are commonly written as $F(\xi)$, $P(\xi)$ and $p(\xi)$. The distribution function of a random variable x has the following properties:

(i) $\lim\limits_{\xi \to -\infty} F(\xi) = 0;\ \lim\limits_{\xi \to \infty} F(\xi) = 1$,

(ii) if $\xi_1 \leqslant \xi_2 \Rightarrow F(\xi_1) \leqslant F(\xi_2)$ (monotone continuous function),

(iii) $\lim\limits_{\xi \uparrow \xi_0} F(\xi) = F(\xi_0)$, $F(\xi)$ is left-continuous or

equivalently, if $\mathscr{P}\{x < \xi\} = F\{\xi\}$, it will be a right-continuous non-decreasing function (see also Borowkow [20]).

(ii) *Joint distribution functions*

Experimental studies of discrete materials frequently require the knowledge of two or more random variables. Thus, if x_1, x_2 are two random variables, they are described by their joint probability distribution function:

$$F_{x_1 x_2}(\xi_1, \xi_2) = \mathscr{P}\{x_1 < \xi_1, x_2 < \xi_2\} \tag{1.55}$$

1.3 Probability, random variables and processes

which is a non-decreasing, right-continuous function with respect to each of the two dimensions, i.e.:

$$\left.\begin{array}{l}\text{(i)} \ F_{x_1 x_2}(\xi_1, -\infty) = 0; \ F_{x_1 x_2}(-\infty, \xi_2) = 0, \\ \text{(ii)} \ F_{x_1 x_2}(\infty, \infty) = 1, \\ \text{(iii)} \ F_{x_1 x_2}(\xi_1, \infty) = F_{x_1}(\xi_1); \ F_{x_1 x_2}(\infty, \xi_2) = F_{x_2}(\xi_2).\end{array}\right\} \quad (1.56)$$

The joint probability density is defined by the mixed derivative of $F_{x_1 x_2}$ and is thus given by the non-negative function:

$$p_{x_1 x_2}(\xi_1, \xi_2) = \frac{\partial^2}{\partial \xi_1 \partial \xi_2} F_{x_1 x_2}(\xi_1, \xi_2). \quad (1.57)$$

Writing the inverse of the above relation, or:

$$F_{x_1 x_2}(\xi_1, \xi_2) = \int_{-\infty}^{\xi_1} \int_{-\infty}^{\xi_2} p_{x_1 x_2}(\xi_1', \xi_2') \, d\xi_1' \, d\xi_2' \quad (1.58)$$

and by letting one of the upper limits approach infinity, leads to the distribution function of a single variable, i.e.

$$\int_{-\infty}^{\infty} d\xi_2' \int_{-\infty}^{\xi_1} p_{x_1 x_2}(\xi_1', \xi_2') d\xi_1' = \mathscr{P}\{x_1 < \xi_1\} = F_{x_1}(\xi_1). \quad (1.59)$$

If both the upper limits in (1.58) reach infinity the interpretation will extend over the entire sample-space in two dimensions so that:

$$\int_{-\infty}^{\infty} \int_{-\infty}^{\infty} p_{x_1 x_2}(\xi_1, \xi_2) d\xi_1 d\xi_2 = 1. \quad (1.60)$$

(iii) *Multi-dimensional distributions; random vectors*

The two-dimensional case can be readily extended to the multi-dimensional one. Thus, recalling that in general a real-valued random variable x on $[\Omega, \mathscr{B}]$ is given by Def. 1 as a function from Ω onto $\mathbb{R}^1$, a n-dimensional random vector $\mathbf{x} = (x_1, x_2, \ldots, x_n)$ is a mapping from $[\Omega, \mathscr{B}]$ into $[\mathbb{R}^n, \mathscr{F}^n]$. Here $\mathbb{R}^n$ is the n-dimensional Euclidean space and $\mathscr{F}^n$ the σ-algebra of Borel subsets of $\mathbb{R}^n$. For a given probability measure $\mathscr{P}$ on $[\Omega, \mathscr{B}]$ one can define another probability measure Q for every set $E \in \mathscr{F}^n$ generated by the random vector $\mathbf{x}$, i.e.:

$$Q\{E\} = \mathscr{P}\{\mathbf{x}^{-1}(E)\}. \quad (1.61)$$

For this measure see also Rényi [17], Borowkow [20] and others. A uniquely defined function $F(\xi_1, \xi_2, \ldots, \xi_n)$ will then express the probability of the events $(-\infty, \xi_1) \times (-\infty, \xi_2) \times \ldots \times (-\infty, \xi_n)$ so that the inequalities $x_1 < \xi_1, x_2 < \xi_2, \ldots, x_n < \xi_n$ are simultaneously satisfied. The function $F(\xi_1, \xi_2, \ldots, \xi_n) = \mathscr{P}\{x_1 < \xi_1, x_2 < \xi_2, \ldots, x_n < \xi_n\}$ is then the distribution function of the random vector or the joint distribution function of the random variables $x_1, x_2, \ldots, x_n$. It has the following properties:

(i) $F_{x_1 x_2 \ldots x_n}(\xi_1, \xi_2, \ldots, \xi_n)$ is a monotonically non-decreasing left-continuous function in each variable.

(ii) $F_{x_1 x_2 \ldots x_n}(\xi_1, \xi_2, \ldots, \xi_n) = F_{x_{i(1)} x_{i(2)} \ldots x_{i(n)}}(\xi_{i(1)}, \xi_{i(2)}, \ldots, \xi_{i(n)})$
where $i(1), i(2), \ldots, i(n)$ is any permutation of $(1, \ldots, n)$.

(iii) $\lim_{x_n \to \infty} F_{x_1 x_2 \ldots x_n}(\xi_1, \xi_2, \ldots, \xi_n) = F_{x_1 x_2 \ldots x_{n-1}}(\xi_1, \xi_1, \ldots, \xi_{n-1})$.

(iv) $\lim_{x_i \to -\infty} F_{x_1 x_2 \ldots x_n}(\xi_1, \xi_2, \ldots, \xi_n) = 0$.

$\lim_{x_i \to \infty} F_{x_1 x_2 \ldots x_n}(\xi_1, \xi_2, \ldots, \xi_n) = 1$.

If the measure $Q(B)$ in relation (1.61) is absolutely continuous with respect to the n-dimensional Lebesgue measure, the distribution function can then be expressed by:

$$F(\xi_1, \xi_2, \ldots, \xi_n) = \int_{-\infty}^{\xi_1} \int_{-\infty}^{\xi_2} \ldots \int_{-\infty}^{\xi_n} p(\xi'_1, \xi'_2, \ldots, \xi'_n) d\xi'_1 d\xi'_2 \ldots d\xi'_n \quad (1.62)$$

in which the probability density is given by:

$$p(\xi_1, \xi_2, \ldots, \xi_n) = \frac{\partial^n F}{\partial \xi_1 \partial \xi_2 \ldots \partial \xi_n} \quad (1.63)$$

and where the derivatives exist almost everywhere in $\mathbb{R}^n$. Hence by condition (i) the probability density $p(\xi_1, \xi_2, \ldots, \xi_n)$ is a measurable non-negative function. Consequently the following definition of a random vector can be given:

Def. 2: "If $\mathbf{x} = (x_1, x_2, \ldots, x_n)$ is an n-dimensional random vector in $\mathbb{R}^n$ the measure Q defined for $E \in \mathscr{F}^n$ in (1.61) is the joint probability and F the joint distribution function. The function $p(\ldots)$ defined by (1.63) is then the joint density function of the random variables $x_1, x_2, \ldots, x_n$."

By considering a fully conditional probability space in the sense of Rényi [17], which is generated by the σ-finite measure $\mathscr{P}$ on the σ-algebra $\mathscr{F}$, one can also define corresponding conditional distribution and density functions. Thus, the conditional distribution function of two real random variables x_1, x_2 designated by $F_{x_1|x_2}(\xi_1|\xi_2)$ or briefly by $F(\xi_1|\xi_2)$ is defined by:

1.3 Probability, random variables and processes

Def. 3:

$$F_{x_1|x_2}(\xi_1|\xi_2) = F(\xi_1|\xi_2) = \mathscr{P}\{x_1 < \xi_1 | x_2 < \xi_2\} \tag{1.64}$$

for all $\xi_1, \xi_2 \in \mathbb{R}^1$

where

$$\mathscr{P}\{x_1 < \xi_1 | x_2 < \xi_2\} = \frac{\mathscr{P}\{x_1 < \xi_1, x_2 < \xi_2\}}{\mathscr{P}\{x_2 < \xi_2\}} = \frac{F(\xi_1, \xi_2)}{F(\xi_2)}. \tag{1.65}$$

Similarly, for two random vectors $\mathbf{x}_1, \mathbf{x}_2$ the conditional distribution function is given by:

$$F_{\mathbf{x}_1|\mathbf{x}_2}(\xi_1|\xi_2) = F(\xi_1|\xi_2) = \mathscr{P}\{\mathbf{x}_1 < \xi_1 | \mathbf{x}_2 < \xi_2\}. \tag{1.66}$$

Correspondingly, the conditional density function of two random variables can be expressed by:

$$p_{x_1|x_2}(\xi_1|\xi_2) = p(\xi_1|\xi_2) = \frac{p(\xi_1, \xi_2)}{p(\xi_2)} \tag{1.67}$$

and, in general, by the chain rule of conditional probabilities:

$$p(\xi_1, \xi_2, \ldots, \xi_n) = p(\xi_n) \prod_{i=1}^{n-1} p(\xi_i | \xi_{i+1}, \ldots, \xi_n). \tag{1.68}$$

In the case of a random vector $(x_1, x_2, \ldots, x_n)$ which has a normal distribution, the probability density becomes:

$$p(\xi_1, \xi_2, \ldots, \xi_n) = \frac{\sqrt{|A|}}{(2\pi)^{n/2}} e^{-\frac{1}{2}Q(\xi_1, \xi_2, \ldots, \xi_n)} \tag{1.69}$$

where $Q = \sum_{i,j=1}^{n} a_{ij} \xi_i \xi_j$

is a positive definite quadratic form and $|A|$ the determinant of the matrix $\|a_{ij}\|$. It can also be shown that:

$$\int_{\mathbb{R}^n} p(\xi_1, \xi_2, \ldots, \xi_n) \, d\xi_1 \, d\xi_2 \ldots d\xi_n = 1. \tag{1.70}$$

(iv) *Some characteristic functions of probability distributions*

Considering a discrete directly given random variable x in $[\mathscr{X}, \mathscr{F}, \mathscr{P}]$ that takes on only a finite number of distinct values, its probability distribution can be stated in terms of a sequence of observable values, i.e.

$$p_k = \mathscr{P}\{x = \xi_k\}, \quad (k = 1, 2, \ldots m). \tag{1.71}$$

Heuristically, one can expect the arithmetic mean or average value of the

Mathematical Preliminaries

observed values of x to be very near to:

$$E(x) = p_1\xi_1 + p_2\xi_2 + \ldots + p_m\xi_m \tag{1.72}$$

which in accordance with the given frequency interpretation of probability (section 1.3.1) would be obtained for n-experiments, where each ξ_k becomes equal to np_k. A more rigorous definition is as follows:

Def. 4: "The expectation of a directly given random variable x on $[\mathcal{X}, \mathcal{F}, \mathcal{P}]$ is defined as the integral of x over $\mathcal{X}$ with respect to the measure $\mathcal{P}$, i.e.

$$E(x) = \int_{\mathcal{X}} x\, d\mathcal{P}. \tag{1.73}$$

if this Lebesgue integral exists."

This conforms with the assumption, that for a large number of observations of a random variable, the mean of the observed values would be $E(x)$ or near to the value of this integral. From the properties of the integral in (1.73) follows immediately that:

(i) the expectation is a linear operation, i.e. for two random variables on the same probability space one has:

$$E(\alpha x + \beta y) = \alpha E(x) + \beta E(y)$$

where α, β are real constants;
(ii) if x is a non-negative random variable, then $E(x) \geq 0$;
(iii) if x is bounded, then $E(x)$ exists.

It is to be noted, that the expected value of a discrete random variable, that takes on infinitely distinct values ξ_n with probabilities $P_n = \mathcal{P}\{x = \xi_n\}$, $n = 1, 2, \ldots$ is given by:

$$E(x) = \sum_{n=1}^{\infty} P_n \xi_n \tag{1.74}$$

if the series converges absolutely. If it converges only conditionally (see Rényi [17]), $E(x)$ is not defined. This can be recognized by the fact that only a random variable, which is constant with probability one ($P\{x = c\} = 1 \Rightarrow E(x) = c$), has a centered distribution. In general, there will be positive and negative deviations, so that for the determination of the distribution and to assess the deviations from the mean value of x, one uses the concept of variance. The latter is defined by:

$$D^2(x) = E\{[x - E(x)]^2\} \tag{1.75}$$

if this quantity exists. If it does not, x is said to have an infinitely large variance. The square root of $D^2(x)$ is referred to as the standard deviation of x or $(D(x))$. These two parameters, i.e. the expectation and variance, are the principal characteristics of the distribution of the random variable x.

1.3 Probability, random variables and processes

Some remarks concerning the basic types of distributions may be indicated here. Thus simple examples of discrete (jump) and continuous distribution functions are given by the Poisson and Gauss distributions. The Poisson distribution is characterized by:

$$\mathscr{P}\{x(\omega) = n\} = P_x(\xi) = \frac{\lambda^n}{n!}e^{-\lambda}, \quad n = 0, 1, \ldots \qquad (1.76)$$

for some $\lambda > 0$. The corresponding distribution function is then:

$$F(\xi) = \sum_{n \leq \xi} \frac{\lambda^n}{n!}e^{-\lambda} \quad \text{for} \quad \xi > 0; \quad F(\xi) = 0 \quad \text{for} \quad \xi < 0. \qquad (1.77)$$

The distribution will be discussed later in dealing with the probabilistic mechanics of discrete fluids. The Gaussian or normal distribution function is given by:

$$F(\xi) = \frac{1}{\sigma\sqrt{2\pi}} \int_{-\infty}^{\xi} e^{-\{(u-a)^2/2\sigma^2\}} \, du \qquad (1.78)$$

in which a is an arbitrary and σ a positive number. In particular, if $a = 0$ and $\sigma = 1$ the distribution function is referred to as a standard Gaussian or normal distribution. Thus for a normally distributed random variable x with parameters a, σ in which a is the mean value and σ its variance, i.e.:

$$\left.\begin{array}{l} E(x) = \int \xi P(d\xi) = \int \xi \, dF(\xi) = a, \\[6pt] \sigma^2(x) = E\{[x - E(x)]^2\}, \end{array}\right\} \qquad (1.79)$$

the standard normal distribution function is:

$$\phi(\xi) = \int_{-\infty}^{\xi} \frac{1}{\sqrt{2\pi}} e^{-(u^2/2)} du \qquad (1.80)$$

where the transformation $u = (\xi - a)/\sigma$ has been used (see also Prohorov and Rozanov [19]). It is seen that Poisson distributions are a one parameter family of distributions in λ and Gaussian distributions depend on the two parameters a and σ defined above. The Gaussian distributions are extensively used in the application of the probabilistic mechanics theory of solids [36].

In certain cases the expectation and variance information about the distribution may not be sufficient and thus higher moments or the expectations of high powers of a random variable are used. These are defined by:

$$M_n(x) = E(x^n) = \int_{-\infty}^{\infty} x^n \, dF(\xi); \quad (n = 1, 2, \ldots). \qquad (1.81)$$

Mathematical Preliminaries

A related definition to (1.81) is that of a central moment of order n, i.e.:

$$m_n(x) = E\{[x - E(x)]^n\}. \tag{1.82}$$

It is apparent that $M_1(x) = E(x)$, $m_1(x) = 0$, $m_2(x) = D^2(x)$. For a bounded random variable all moments exist and the distribution is uniquely determined. By using the Fourier–Stieltjes transform of the distribution function $F(\xi)$ or:

$$\varphi_x(t) = \int_{-\infty}^{\infty} e^{i\xi t}\, dF(\xi); \quad (-\infty < t < \infty) \tag{1.83}$$

one obtains the characteristic function of the random variable in probability theory. If the distribution is continuous with the probability density function $p(\xi)$ then:

$$\varphi_x(t) = \int_{-\infty}^{\infty} e^{i\xi t} p(\xi)\, d\xi. \tag{1.84}$$

Since by definition the Lebesgue integral of any real Borel measurable function $g(x)$ of the real variable x is given by:

$$E\{g(x)\} = \int_{-\infty}^{\infty} g(\xi)\, dF(\xi) \tag{1.85}$$

one can express (1.83) also by:

$$\varphi_x(t) = E\{e^{ixt}\}. \tag{1.86}$$

The moments sometimes required for a more complete description of the distribution function of a bounded random variable can be obtained in terms of the characteristic function by taking the derivatives of the function at $t = 0$. Thus

$$M_n(x) = i^{-n} \varphi_x^{(n)}(0); \quad (n = 1, 2, \ldots). \tag{1.87}$$

Assuming that all moments of a certain distribution are given and by taking the Laplace transform of the distribution function $F(\xi)$, i.e.:

$$\left. \begin{array}{l} \psi_x(s) = \displaystyle\int_{-\infty}^{\infty} e^{\xi s}\, dF(\xi) = \int_{-\infty}^{\infty} e^{\xi s} p(\xi)\, d\xi \\[2mm] = 1 + \displaystyle\sum_{n=1}^{\infty} \frac{M_n s^n}{n!} \end{array} \right\} \tag{1.88}$$

in which for a bounded random variable, the series in (1.88) converges. The

1.3 Probability, random variables and processes

function $\psi_x(s)$ is referred to as moment generating function of the random variable x. The notion of a characteristic function can also be applied to random vectors. Thus, if $\mathbf{x} = (x_1, x_2, \ldots, x_n)$ is an n-dimensional random vector, its characteristic function $\varphi_\mathbf{x}(t_1, t_2, \ldots, t_n)$ for any real values $t_1, t_2, \ldots, t_n$ is given by:

$$\varphi_\mathbf{x}(t) = E\{e^{i(\mathbf{x},t)}\} = \int e^{i(\mathbf{x},t)} F_{x_1, x_2, \ldots, x_n}(\mathrm{d}\xi_1, \mathrm{d}\xi_2, \ldots, \mathrm{d}\xi_n) \qquad (1.89)$$

where t denotes the vector $(t_1, t_2, \ldots, t_n) \in \mathbb{R}^n$ and (x, t) is the scalar product of the vectors x and t, i.e.:

$$(x, t) = \sum_{k=1}^{n} x_k t_k. \qquad (1.90)$$

For a more comprehensive discussion on the concepts of random variables, their distribution functions, characteristic functions, etc., the reader is referred to Borowkow [20], Rényi [17], Neveu [21] and others [22, 23]. In this context considerations of the determination of distributions of random variables, their independence and some general theorems associated with this notion are considered below.

(v) *Independent random variables*

The random variables $x_1, x_2, \ldots, x_n$ are called stochastically independent, if

$$\mathscr{P}\{x_1 \in B_1, \ldots, x_n \in B_n\} = \mathscr{P}\{x_1 \in B_1\} \ldots \mathscr{P}\{x_n \in B_n\} \qquad (1.91)$$

where $B_1, \ldots, B_n$ are arbitrary Borel sets of the real line $\mathbb{R}^1$. Thus, if F_k denotes the distribution function of x_k $(k = 1, 2, \ldots, n)$ the joint distribution function of the random variables $x_1, x_2, \ldots, x_n$ is given by:

$$F(x_1, x_2, \ldots, x_n) = F_1(x_1) F_2(x_2) \ldots F_n(x_n). \qquad (1.92)$$

If this distribution is absolutely continuous with the density function $p_k(\xi)$, $(k = 1, 2, \ldots, n)$ then the density function of the joint distribution of the random variables is:

$$p(\xi_1, \xi_2, \ldots, \xi_n) = p_1(\xi_1) p_2(\xi_2) \ldots p_n(\xi_n). \qquad (1.93)$$

These statements can be extended to vector-valued random variables. Thus, let $x_1, x_2, \ldots, x_n$ be independent real or vector-valued random variables and $g_1(\xi), g_2(\xi), \ldots, g_n(\xi)$ Borel-measurable real or vector valued functions such that, if x_j is an r_j-dimensional vector-valued random variable $(r_j \geq 1)$ then g_j is a function of r_j variables. The random variables $g_1(x_1), g_2(x_2), \ldots, g_n(x_n)$ are then independent. It is of further interest to note that for two independent

32 Mathematical Preliminaries

variables x, y having finite expectations, one has:

$$E\{x,y\} = E\{x\}E\{y\}. \qquad (1.94)$$

Two random variables x and y for which $E\{x\}$, $E\{y\}$ and $E\{x, y\}$ exist and (1.94) is valid, are said to be uncorrelated. Hence independent random variables having finite expectations are uncorrelated. In the case of two random variables with finite expectations and variances, one can define a correlation coefficient $\rho(x, y)$ as follows:

$$\rho(x,y) = \frac{E\{x,y\} - E\{x\}E\{y\}}{D\{x\}D\{y\}}. \qquad (1.95)$$

If $\rho(x, y) = 0$ the random variables x and y are uncorrelated (see also Pugachev [23], Borowkow [20], Rényi [17] and others). An obvious generalization of the statement in (1.94) is that, if $x_1, x_2, \ldots, x_n$ are independent random variables with finite expectations, then $x_1, x_2, \ldots, x_n$ have also finite expectation or:

$$E\{x_1, x_2, \ldots, x_n\} = \prod_{k=1}^{n} E\{x_k\}. \qquad (1.96)$$

In particular a detailed discussion of sequences of independent random variables can also be found, for instance, in Rosenblatt [24].

The concept of independence of random variables is closely related to the independence of σ-algebras. Thus, considering the event structure in a probability space $[\mathscr{X}, \mathscr{F}, \mathscr{P}]$ and two classes of events $\mathscr{E}_1, \mathscr{E}_2$ contained in the σ-algebra $\mathscr{F}$ of the space $\mathscr{X}$, the two classes are called independent, if for all events $E_1 \in \mathscr{E}_1$ and $E_2 \in \mathscr{E}_2$ the following relation holds:

$$\mathscr{P}\{E_1 E_2\} = \mathscr{P}\{E_1\}\mathscr{P}\{E_2\} \qquad (1.97)$$

(see also section 1.3.1). In a sequence of classes of events denoted by $\{\mathscr{E}_n\}_{n=1}^{\infty}$ the events will be independent, if for any choice of the integer n:

$$\mathscr{P}\left\{\bigcap_{i=1}^{k} E_{n_i}\right\} = \prod_{i=1}^{k} \mathscr{P}\{E_{n_i}\} \qquad (1.98)$$

for arbitrary $E_{n_i} \in \mathscr{E}_i$. Thus, the sub-algebras of independent algebras of events are also independent. In this context an approximation theorem by Borowkow in relation to directly given random variables states that:

"if $[\mathscr{X}, \mathscr{F}, \mathscr{P}]$ is a probability space and $\mathscr{E}$ the σ-algebra of a certain class of events, then there exists for each $E \in \mathscr{F}$ a sequence $E_n \in \mathscr{E}$" such that:

$$\left.\begin{array}{l} \lim_{n \to \infty} \mathscr{P}\{E_n E' \cup E'_n E\} = 0 \\[4pt] \text{or equivalently:} \\[4pt] \lim_{n \to \infty} \mathscr{P}\{E \setminus E_n\} = \lim_{n \to \infty} \mathscr{P}\{E_n \setminus E\} = 0. \end{array}\right\} \qquad (1.99)$$

1.4 Basic stochastic processes

Since $\mathscr{P}\{E\} = \mathscr{P}\{E_n E\} + \mathscr{P}\{E'_n E\} = \mathscr{P}\{E_n\} - \mathscr{P}\{E_n E'\} + \mathscr{P}\{E'_n E\}$ it follows that:

$$\mathscr{P}\{E\} = \lim_{n \to \infty} \mathscr{P}\{E_n\} \tag{1.100}$$

(see also Halmos [11]). Hence, by denoting the smallest σ-subalgebra of $\mathscr{F}$ by $\sigma(x)$ with respect to which the mapping of x is still measurable, it may be stated that the random variables $x_1, x_2, \ldots, x_n$ will be independent, if the subalgebras $\sigma(x_1), \sigma(x_2), \ldots, \sigma(x_n)$ are independent. Analogously, the independence of random vectors will be based on:

$$\mathscr{P}\{x_1 \in B_1, x_2 \in B_2, \ldots, x_n \in B_n\} = \Pi \mathscr{P}\{x_j \in B_j\} \tag{1.101}$$

where the B_j's represent Borel sets in the subspaces of corresponding dimensions. Hence, it is always possible to define a finite sequence of independent random variables by considering the random quantities $x_1, x_2, \ldots, x_n$ in the space of elementary events $\mathbb{R}^1 \times \mathbb{R}^1 \ldots n\text{-times} = \mathbb{R}^n$ and the corresponding σ-algebras $\mathscr{B}_1 \times \mathscr{B}_2 \times \ldots \times \mathscr{B}_n = \mathscr{B}^n$ induced by the Borel sets $B_1 \times B_2 \times \ldots \times B_n \subset \mathbb{R}^n$.

In the infinite dimensional case this is somewhat more difficult and one has to introduce Kolmogorov's extension theorem (see, for instance, Prohorov and Rozanov [19]). This theorem states, that on the basis of compatible distributions $\mathscr{P}_n$ on the finite dimensional space $\mathbb{R}^n$, it is possible to define a unique probability measure $\mathscr{P}_\infty$ on $[R^\infty, \mathscr{B}^\infty]$ such that each $\mathscr{P}_n$ is the projection of $\mathscr{P}^\infty$ on $\mathbb{R}^n$.

1.4 BASIC STOCHASTIC PROCESSES

In the development of probabilistic mechanics the notion of a random function or random process is fundamental. In the preceding sections random variables and certain relations of the theory of probability were briefly considered. Such variables assume certain values in any particular trial or random experiment. In stochastic mechanics, however, random variables can assume in the outcome of an experiment a set of values, which will depend on time as well as on other parameters. One deals therefore with a function that has only one uniquely specified value corresponding to each $\omega \in \Omega$ and instant of time over an interval in which actual observations are made. In other words, if another observation is made under identical physical conditions, a different function would be obtained. Such functions are called stochastic functions and can be designated by $x(t, \omega)$, $t \in T$, $\omega \in \Omega$. Usually a function of this type for a specific ω is referred to as a realization or sample function $x_t(\omega)$. In the context of the earlier definition concerning directly given random variables, it is to be noted that for each elementary outcome $\omega \in \Omega$ the function $x(t, \omega)$ with $t \in T$ having values in $[\mathscr{X}, \mathscr{F}]$ is referred to as a realization of the random process

34 Mathematical Preliminaries

$x = x(t)$. Thus $x(t)$ is called directly given, if each ω is described by the corresponding realization $x = x(t)$ in the probabilistic function space $\mathscr{X}^T$ of all functions on the time parameter set T with values in $[\mathscr{X}, \mathscr{F}]$ where from before $\Omega = \mathscr{X}$ (see Prohorov and Rozanov [19]). Hence, the random function may be regarded as specified, if for each $t \in T$ its distribution is given by:

$$F_t(\xi) = F_x(\xi, t) = \mathscr{P}\{x(t) < \xi\}, t \in T. \qquad (1.102)$$

Thus considering the two-dimensional case, one can write for $x(t_1, t_2) = [x(t_1), x(t_2)]$ the joint distribution function as follows:

$$F_2(x_1, x_2; t_1 t_2) = F_{t_1, t_2}(x_1, x_2) = \mathscr{P}\{x(t_1) < \xi_1, x(t_2) < \xi_2\}. \qquad (1.103)$$

This is readily extended to the n-dimensional case, where

$$x(t_1, t_2, \ldots, t_n) = [x(t_1), x(t_2), \ldots, x(t_n)]$$

with the corresponding n-dimensional distribution function:

$$F_k(x_1, x_2, \ldots, x_k; t_1, t_2, \ldots, t_k)$$

$$= \int_{-\infty}^{\infty} {}_{(n-k)} \int_{-\infty}^{\infty} F_n(x_1, x_2, \ldots, x_n; t_1, t_2, \ldots, t_n) dx_{k+1} \ldots dx_n \qquad (1.104)$$

for all $n > k$ (see also Yaglom [30]). The distribution function as stated in (1.103), for instance, can be shown to satisfy the symmetry condition and compatibility condition for any set $t_{k+1}, \ldots, t_n$ provided $k < n$. The set $t \in T$ consists frequently of the entire set of real numbers going from $-\infty < t < \infty$ or for the half-line $t > 0$. The family $\{x(t), t \in T\}$ is then considered as a time-continuous random process. If T is a set of integers only, one has a random sequence encountered in the analysis of temporal discrete systems. A distinct property of a random function is its stationarity, i.e. that all finite dimensional distribution functions that define the process $x(t)$ remain the same, if the set $t \in T, t_1, \ldots, t_n$ is shifted along the time axis. Thus, for all stationary random functions the one-dimensional distribution is the same, whilst the two-dimensional distribution depends only on the difference $t_2 - t_1$, etc. The concept of a stochastic function briefly outlined above can be readily extended to vectors and tensors. Such quantities are contained in statistical field theories as discussed in subsequent chapters of this text. Thus, a random tensor, for example, may be regarded as one whose elements are stochastic variables or stochastic functions. In general, any mathematical quantity in the deterministic form has its counterpart in the stochastic description. For a more comprehensive study the reader is referred to the texts given in the Bibliography [23, 25, 26]. In the application of random functions in probabilistic mechanics various theorems are required, in particular those relating to convergence, continuity, the definition of stationarity of a random process, etc. These characteristics of

1.4 Basic stochastic processes 35

random functions become important when dealing with the differentiation and integration of random field variables.

(i) *Some characteristics of stochastic functions*

From the point of view of probability theory by considering a sequence of random variables $x_1, x_2, \ldots, x_n$ the following definitions for convergence can be given (see also Fréchet [27] and [28]):

(a) *Convergence everywhere* (e-convergence):

$$x_n(\omega) \to x(\omega), \quad \text{if} \quad \lim_{n \to \infty} x_n(\omega) = x(\omega) \quad \text{for} \quad \forall \omega \in \Omega. \tag{1.105}$$

(b) *Convergence almost everywhere* (a.e.-convergence):
$(x_1 \ldots x_n, \ldots)$ converges a.e. to x as $n \to \infty$ if the probability of every sequence of realized values of x_n converging to x is one, i.e.:

$$\mathscr{P}\{\lim_{n \to \infty} x_n = x\} = 1; \quad \lim_{n \to \infty} x_n = x \text{ (a.e.).} \tag{1.106}$$

(c) *Convergence in probability* (p-convergence):
x_n converges in probability to x as $n \to \infty$ if for every $\varepsilon > 0$

$$\lim_{n \to \infty} \mathscr{P}\{|x_n - x| > \varepsilon\} = 0, \quad \forall \varepsilon > 0. \tag{1.107}$$

(d) *Convergence in quadratic mean* (mean square convergence, m.s.):
x_n converges to x in the m.s. as $n \to \infty$ if
$\lim E\{|x_n - x|^2\} = 0$ or symbolically

$$\lim_{n \to \infty} x_n = x \quad \text{by definition, if} \quad \lim_{n \to \infty} E\{|x_n - x|^2\} = 0. \tag{1.108}$$

(e) *Convergence in distribution*:
x_n converges to x as $n \to \infty$ if

$$\lim_{n \to \infty} F(x_n) = F(x) \tag{1.109}$$

at every point of the continuous distribution function $F(x)$.

These convergence definitions are not equivalent, e.g. it is possible for a sequence of random variables to converge almost everywhere, but not in the mean square. These definitions can, however, be extended to stochastic functions. Thus, for instance:

$$\lim_{t \to t_0} x(t) = x(t_0), \quad \text{if} \quad \lim_{t \to t_0} E\{|x(t) - x(t_0)|^2\} = 0. \tag{1.110}$$

36 Mathematical Preliminaries

For a more comprehensive study of convergence theorems see, for instance, Fréchet [27], and others. Other characteristics of random functions of importance are regularity and continuity. Thus, random functions which have a finite mean square are called regular random functions. At points or sets of points, where this criterion does not hold, one speaks of m.s. singularities of the function $x(t)$. A random function $x(t)$ with realizations at $x(t)$ and $x(t+s)$, $s, t \in T$ is said to be continuous at point t in probability, if

$$\left.\begin{array}{l}\lim_{s \to 0} \mathscr{P}\{|x(t+s)-x(t)| \leqslant \varepsilon\} = 1 \quad \text{for} \quad \forall \varepsilon > 0 \\ \text{or equivalently, if } \lim_{s \to 0} \mathscr{P}\{|x(t+s)-x(t)| > \varepsilon\} = 0 \text{ for } \forall \varepsilon > 0.\end{array}\right\} \quad (1.111)$$

A random function is continuous in the m.s. sense, if

$$\lim_{s \to 0} E\{|x(t+s)-x(t)|^2\} = 0. \tag{1.112}$$

Apart from the above given definitions stochastic integrals are also defined in the mean square sense, i.e.:

(i) Considering the stochastic function $x(t)$ on an interval $(a, b) \subset T$ subdividing it into arbitrary intervals Δt_k and taking

$$\text{the sum } \sum_{k=1}^{n} x_k \Delta t_k,$$

where x_k is the random variable associated with an arbitrary point in Δt_k and if

$$\text{l.i.m.} \sum_{k=1}^{n} x_k \Delta t_k = \int_{a}^{b} x(t)\,dt \tag{1.113}$$

exists, $x(t)$ is said to have a stochastic Riemann integral on (a, b).

(ii) Defining a suitable measure $\mathscr{P}(A)$ on the time-parameter space, where $\mathscr{P}(A)$ is a non-negative set function defined for all Borel sets of the T-space, then $x(t)$ can be considered on the set A of finite Lebesgue measures. Dividing A into n-disjoint sets A_k and forming

$$\sum_{k=1}^{n} x_k \mathscr{P}\{A_k\}$$

(x_k is here the random variable associated with a point in A_k) and if,

$$\text{l.i.m.} \sum_{k=1}^{n} x_k \mathscr{P}\{A_k\} = \int_{A} x(t)\,d\mathscr{P}(t) \tag{1.114}$$

exists, $x(t)$ is said to have a stochastic Lebesgue integral on A.

(iii) Similarly, the stochastic Stieltjes integral of $x(t)$ with respect to a set of random functions $S(A_k)$ on A_k for all Borel sets A_k of the T-space is defined by:

$$\underset{\substack{n \to \infty \\ S(A_k) \to 0}}{\text{l.i.m.}} \sum_{k=1}^{n} x_k S(A_k) = \int_{A_k} x(t) \, dS(t). \quad (1.115)$$

Most theorems of ordinary differential and integral calculus apply also to stochastic functions but require certain modifications. The latter originate from probabilistic and measure-theoretical considerations. One concept, however, is new, e.g. the interchangeability of the expectation operator and the differentiation of random quantities. Thus the expectation of a derivative or product of derivative and of a random function $x(t)$ is given by:

$$E\left\{\frac{dx(t)}{dt}\right\} = \frac{d}{dt} E\{x(t)\} \quad (1.116)$$

and

$$E\left\{x(t) \frac{dx(s)}{ds}\right\} = \frac{\partial}{\partial s} E\{x(t)x(s)\} \quad (1.117)$$

where the derivatives are mean square differential operators, i.e.

$$x'(t) = \frac{d}{dt} x(t)$$

if the limit:

$$\lim_{\Delta t \to 0} E\left\{\left|\frac{x(t + \Delta t) - x(t)}{\Delta t} - x'(t)\right|^2\right\} = 0 \text{ exists.} \quad (1.118)$$

A necessary and sufficient condition that $x(t)$ has a m.s. derivative at a point t is that:

$$\frac{\partial^2}{\partial t \, \partial s} E\{x(t)x(s)\} \quad \text{exists at} \quad (t, s) = (t, t).$$

If $x(t)$ has a m.s. derivative for every $t \in T$, then

$$\frac{\partial}{\partial t_1} E\{x(t_1)x(t_2)\} \quad \text{and} \quad \frac{\partial}{\partial t_2} E\{x(t_1)x(t_2)\}$$

exist as well for all $t_1, t_2 \in T$. Using the product moment of $x(t)$, i.e.: $m_{1,2,\ldots,n}(t_1, t_2, \ldots, t_n) = E\{x_1, x_2, \ldots, x_n\}$ relation (1.117) can be generalized to:

$$E\{x_1^{(k_1)} x_2^{(k_2)} \ldots x_n^{(k_n)}\} = \left(\frac{\partial}{\partial t_1}\right)^{k_1} \left(\frac{\partial}{\partial t_2}\right)^{k_2} \ldots \left(\frac{\partial}{\partial t_n}\right)^{k_n} E\{x_1, x_2 \ldots x_n\} \quad (1.119)$$

where $x_i^{(k_i)} = \dfrac{\partial^{(k_i)} x(t_i)}{\partial t_i^{(k_i)}}$.

(ii) Second-order characteristics of stochastic functions

In the analysis of probabilistic mechanics and particularly in the modelling of the response behaviour of discrete materials in which the field quantities are either random variables or random processes, the so-called second-order properties of random functions are of great significance. Analogous to the case of random variables the moment function of order m of the scalar random process $x(t)$ can be written as: $E\{x(t_1)x(t_2)\ldots x(t_n)\}$, which in general is a function of $t_1, t_2, \ldots, t_n$ and is determined by the joint distribution of the values of the process at the n-points selected. The most frequently used functions of stochastic processes are the first and second moments. The first-order moment functions is usually denoted by $m_x(t)$ and is called the mean function of the process:

$$m_x(t) = E\{x(t)\}; \quad t \in T \qquad (1.120)$$

and will exist so long as $E\{x(t)\} < \infty$ for all $t \in T$. It traces a path or trajectory for the mean value of the process $x(t)$ for each t. The second-order moment function is called correlation function of the process $x(t)$, i.e.:

$$E\{x(t_1)x(t_2)\} = R_x(t_1, t_2); \quad t_1, t_2 \in T. \qquad (1.121)$$

It exists, if $E\{|x(t)|^2\} < \infty$ for $\forall t \in T$ since by Schwartz's inequality (see also ref. [7]):

$$E\{x(t_1)x(t_2)\} \leqslant [E\{|x(t_1)|^2\} E\{|x(t_2)|^2\}]^{\frac{1}{2}}. \qquad (1.122)$$

Any random process satisfying $E\{|x(t)|^2\} < \infty$ for all $t \in T$ is said to be a finite variance process. Analogous to the correlation coefficient given in section 1.3.2 (v), for the random functions $x(t_1), x(t_2)$, it becomes:

$$\left.\begin{aligned}\rho_{x(t_1)x(t_2)} &= \frac{E\{[x(t_1) - m_{x(t_1)}][x(t_2) - m_{x(t_2)}]\}}{\sqrt{E\{[x(t_1) - m_{x(t_1)}]^2\} E\{[x(t_2) - m_{x(t_2)}]^2\}}} \\ &= \frac{R_x(t_1, t_2) - m_{x(t_1)} m_{x(t_2)}}{\sigma_{x(t_1)} \sigma_{x(t_2)}}\end{aligned}\right\} \qquad (1.123)$$

in which $\sigma_{x(t)}$ is the standard deviation of $x(t)$. If the process is normalized so that $m_{x(t)} = 0$ and $\sigma^2_{x(t)} = 1$ for each $t \in T$, the correlation coefficient is equal to the correlation function at any two time instants t_1, t_2 (see also Pugachev [23]):

$$\rho_{x(t_1)x(t_2)} = R_x(t_1, t_2). \qquad (1.124)$$

The two functions, i.e. the mean and correlation functions, can be used to describe the variance of the value of the process at any given time t. Evidently one can define the covariance between the values at any two points t_1 and t_2 as follows:

$$\left.\begin{aligned}R_x(t, t) &= E\{|x(t)|^2\} \to \sigma^2_{x(t)} = R_x(t, t) - m^2_{x(t)} \\ \text{so that the covariance function becomes:} & \\ \text{Cov}\{x(t_1)x(t_2)\} &= R_x(t_1, t_2) - m_x(t_1) m_x(t_2)\end{aligned}\right\} \qquad (1.125)$$

1.4 Basic stochastic processes

where $\sigma_x^2(t)$, $t \in T$ is the variance function of the process. By using the Schwartz inequality again, it can be shown that if $R_x(t_1, t_2)$ is the correlation function of the real-valued random process $x(t)$, $t \in T$, it must be symmetric and non-negative definite (see also Khintchine [29], Yaglom [30]). It is to be noted that in the usual terminology concerning stochastic processes, the function $R_x(t_1, t_2)$ is called auto-correlation function to indicate that the two random variables considered, i.e. $x(t_1)$ and $x(t_2)$ belong to the same process. Thus, if two different processes are analyzed the term cross-correlation function is employed, where:

$$R_{xy}(t_1, t_2) = E\{x(t_1)y(t_2)\} \tag{1.126}$$

in which $x(t_1)$ and $x(t_2)$ belong to different (perhaps related) random processes. Further one uses frequently the term cumulant functions, where for instance:

$$\left.\begin{aligned}K_1\{x(t_1)\} &= E\{x(t_1)\} = m_x(t_1),\\ K_2\{x(t_1)x(t_2)\} &= E\{[x(t_1) - m_x(t_1)][x(t_2) - m_x(t_2)]\}.\end{aligned}\right\} \tag{1.127}$$

Whilst higher-order cumulant functions are not easily identified, the nth cumulant function can be expressed in terms of the nth order and lower-order moment functions. In certain applications it may be more convenient to employ cumulant functions rather than moment functions. The second cumulant function, for instance, given by:

$$K_x(t_1, t_2) = K_2\{x(t_1)x(t_2)\} \tag{1.128}$$

is called auto-covariance function. Analogously, if two random processes are considered one has a cross-covariance function given by:

$$\left.\begin{aligned}K_{xy}(t_1, t_2) &= E\{[x(t_1) - m_x(t_1)][y(t_2) - m_y(t_2)]\}\\ &= R_{xy}(t_1, t_2) - m_x(t_1)m_y(t_2).\end{aligned}\right\} \tag{1.129}$$

The importance of the first- and second-order statistical properties of a random process becomes apparent, when dealing with a process in which the higher characteristics can be established, if the first and second ones are known. This is the case for the Gaussian random process which occurs in many physical phenomena. Moreover, by employing the well-known Chebyshev inequality pertaining to a random variable, i.e.:

$$\mathscr{P}\{|x - m_x| \geqslant \varepsilon \sigma_x\} \leqslant \frac{1}{\varepsilon^2}, \quad \varepsilon > 0 \tag{1.130}$$

an upper bound for the probability of the event $|x(t) - m_x(t)| \geqslant \varepsilon$ at any point $t \in T$ can be formed from the mean and variance functions of the process $x(t)$. It is equally possible to assess the probability of such an event for any t in the closed interval $a \leqslant t \leqslant b$ (see, for instance, Parzen [31]).

40 Mathematical Preliminaries

For an arbitrary function $h(t)$ the auto-correlation function satisfies the following inequality:

$$R_{xx}(t_i, t_j) h(t_i) h^*(t_j) \geq 0 \qquad (1.131)$$

where a repeated index means summation and i, j is the range from 1 to any finite integer n. The asterisk indicates the complex conjugate of the function. This relation also holds for complex-valued random processes in which the auto-correlation function is defined by:

$$R_{xx}(t_1, t_2) = E\{x(t_1) x^*(t_2)\}. \qquad (1.132)$$

The symmetry property in this case becomes hermitian, i.e. $R_{xx}(t_1, t_2) = R_{xx}^*(t_2, t_1)$. A theorem by Bochner [32] asserts that every non-negative definite function satisfying (1.131) has a non-negative Fourier transform. Thus the Fourier transform of $R_{xx}(\tau)$ is given by:

$$S(\omega) = \frac{1}{2\pi} \int_{-\infty}^{\infty} R_{xx}(\tau) \exp[-i\omega\tau] \, d\tau \geq 0. \qquad (1.133)$$

This function is of utmost importance in the theory of stationary random processes (see also Doob [33], Yaglom [30] and others). $S(\omega)$ is called the mean square spectral density or briefly spectral density of the stationary process $x(t)$ and is evidently a non-negative function. By the Fourier inversion theorem one has:

$$R(\tau) = \int_{-\infty}^{\infty} e^{i\omega\tau} S(\omega) \, d\omega. \qquad (1.134)$$

The relations (1.133, 1.134) are known as the Wiener–Khintchin theorem for stochastic processes and form the basis for a generalized harmonic analysis (Wiener [34]). It permits the application of spectral analysis to be extended to random functions. The spectral analysis becomes important, when dealing with molecular dynamics of discrete fluids as will be discussed in Chapter 5 of this text.

(iii) *Gaussian- and Poisson-type stochastic processes*

Since a large number of physical phenomena can be modelled by Gaussian random processes, the latter are briefly mentioned here. A Gaussian process is one where all distribution functions are joint normal distributions. Thus, in terms of the nth order probability density function the process is described by:

$$p_n(x_1, x_2, \ldots, x_n; t_1, t_2, \ldots, t_n)$$

$$= \frac{\Delta^{\frac{1}{2}}}{(2\pi)^{n/2}} \exp\left[-\frac{1}{2} \sum_{j,k} \Delta_{jk} (x_j - m_j)(x_k - m_k)\right] \qquad (1.135)$$

where $m_j = m(t_j)$ are the means, $\Delta = |K_{jk}|$ the determinant of the auto-covariance function K_{jk} and Δ_{jk} the cofactor of K_{jk} in the determinant. By using a joint characteristic functional $\phi_x(t)$ of the jointly distributed Gaussian random variables $x(t_1)$, $x(t_2)$, ..., $x(t_n)$ (see 1.83) of the form:

$$\phi_x(t) = \exp\left[i \sum_{j=1}^n \varphi_j m_x(t_j) - \tfrac{1}{2} \sum_{j,k=1}^n \sum K_{xx}(t_j, t_k)\varphi_j \varphi_k\right] \quad (1.136)$$

it is seen that the normal random function is completely specified by its mean $m(t_j)$ and its auto-covariance $K_{jk}(t_j, t_k)$, since from these functions the distribution function at n-points of the process can be obtained. An assignment of different parametric values to a Gaussian process yields again Gaussian random variables. It can be shown by using the consistency theorem of probability (see, for instance, Loève [22]) that conversely, if for any n, $x(t_1)$, $x(t_2)$, ..., $x(t_n)$ are Gaussian random variables, the process $x(t)$ must be also a Gaussian random process. For an arbitrary weakly stationary random process (Yaglom [30]) $m_x(t)$ is a constant and $K_{xx}(t_1, t_2)$ a function of $(t_1 - t_2)$ only. For a Gaussian process, however, no distinction has to be made between weak and strong stationarity, since one implies the other (see Doob [33]). A linear transformation of a set of Gaussian random variables leads to a new set of Gaussian random variables. Analogously, any linear operation on a Gaussian process results in another Gaussian process. One can generalize such processes by introducing a linear functional on a linear space U, i.e.: $f = \langle u, x \rangle$. The process will be Gaussian, if its characteristic functional ϕ_x is of the form:

$$\phi_x(u) = \exp\left[iE(\langle u, x \rangle) - \tfrac{1}{2}C(u, v)\right] \quad (1.137)$$

where $E(\langle u, v \rangle)$ is the expectation of the generalized process and $C(u, v)$ the scalar product of $u, v \in U$ or the correlation functional (see also Prohorov and Rozanov [19]).

Another class of stochastic processes $x(t), t \in T$ is that of independent increments, if for all $t_1 < t_2 < \ldots < t_k < t_{k+1} \in T$ and every $k = 1, 2, \ldots$ the random variables $x(t_2) - x(t_1)$, $x(t_3) - x(t_2)$, ..., $x(t_{k+1}) - x(t_k)$ are mutually independent (Doob [33]), Feller [28]). Since there exists a simple relationship between $x(t_1), \ldots, x(t_k)$ and their increments, the process will be specified by the density of an increment $x(t) - x(t')$ for all $t' < t$ and the first-order density $p_0(x)$ at some $t_0 \in T$. An important representative of this group is the Poisson process. Thus a random process with independent increments satisfying:

$$P\{x(t) - x(t') = k\} = \frac{[\lambda(t-t')]^k}{k!} e^{-\lambda(t-t')} \quad (1.138)$$

for any $t, t' \in T$, $t' \leq t$ is called a Poisson process. It will be employed in Chapter 5 of this text.

1.5 RANDOM FIELDS

It has been shown in previous work [35, 36] that the evolution of the behaviour of discrete media can be represented in terms of stochastic processes. In the development of probabilistic mechanics the Markov processes to be discussed in Chapter 2 are of special interest since they can be used in the description of a large number of physical phenomena. However, there are certain phenomena in the dynamics of structured solids or in the molecular dynamics of discrete fluids, where it becomes necessary to generalize stochastic processes to random fields. As shown previously, if $[\mathscr{X}, \mathscr{F}, \mathscr{P}]$ is a probabilistic function space, a real-valued stochastic process is a map $x: \mathbb{R}^1 \times \mathscr{X} \to \mathbb{R}^1$ such that $\omega \mapsto x(t, \omega)$ or $x_t(\omega)$ becomes a random variable for all $t \in \mathbb{R}^1$. x is then considered as a random variable to represent the random motion of an element of the microstructure that evolves with time. Hence, for a fixed t the random variable $\omega \mapsto x_t(\omega)$ may be, for example, the position coordinate of the centre of mass of a microelement, whilst for a fixed ω, $t \mapsto x_t(\omega)$ becomes a sample function of this coordinate as it progresses with time. In order to generalize this basic notion consider a space $\mathscr{U}$ to be a real or complex inner product space, where $\mathscr{U}$ is assumed to have a norm topology determined by the inner product. If $[\mathscr{X}, \mathscr{F}, \mathscr{P}]$ is a probability space and a set R of complex-valued random variables on it, the map $\Phi: \mathscr{U} \to R[\mathscr{X}, \mathscr{F}, \mathscr{P}]$ becomes a random functional. Hence, Φ is a stochastic process indexed by $\mathscr{U}$. The random functional Φ is said to be linear if:

$$\left. \begin{array}{l} \Phi(\alpha\varphi + \beta\psi) = \alpha\Phi(\varphi) + \beta\Phi(\Psi) \quad \text{for every} \quad \alpha, \beta \in \mathbb{C} \\ \qquad\qquad\qquad\qquad\qquad\qquad (\mathscr{U} \text{ being a complex space}) \\ \text{or, for every } \alpha, \beta \in \mathbb{R}^1 \\ \qquad\qquad\qquad\qquad\qquad\qquad (\mathscr{U} \text{ being a real space}). \end{array} \right\} \quad (1.139)$$

The random functional is continuous, if $\varphi_i \to \varphi$ in $\mathscr{U}$ which implies that $\Phi(\varphi_i) \to \Phi(\varphi)$ in probability. The above formulation of a linear random functional leads to a definition of a generalized random field in the sense of Yaglom [30], Gel'fand [37]. Thus, a continuous linear random functional is a random field. Thus, if $x: \mathbb{R}^n \times \mathscr{X} \to \mathbb{C}$ is a stochastic process, $\Phi(\mathbf{r}) = x(\mathbf{r}, \cdot)$ is a random functional on the space $\mathbb{R}^n$, if it is a physical space. x in this case could be some field quantity, where the path $\mathbf{r} \mapsto x(\mathbf{r}, \omega_0)$ represents a sample of values of the measurements of x at each point in this space. The random variable $\omega \to x(\mathbf{r}_0, \omega)$ represents the statistical outcome of the values of x at points $\mathbf{r}_0$ in $\mathbb{R}^n$.

Examples of physical spaces of the $\mathbb{R}^n$ type will be considered in dealing with the quasi-statics and dynamics of structured solids for instance or with the molecular dynamics of discrete fluids. If the physical field in the Euclidean space $\mathbb{R}^n$ is characterized in terms of a stochastic process $x_t: \mathbb{R}^n \times \mathscr{X} \to \mathbb{C}$, it corresponds strictly to precise measurements at a point $\mathbf{r}_0 \in \mathbb{R}^n$. However,

1.5 Random fields

measurements due to experimental constraints can only be carried out over a specific range of accuracy, and hence can only be interpreted by the mean value or a weighting function on $\mathbb{R}^n$ with a random variable. By physical considerations such a functional should be linear and continuous and hence represents a random field. In accordance with the definitions for probabilistic quantities given previously one can define for the random functional $\Phi: \mathcal{U} \to R[\mathcal{X}, \mathcal{F}, \mathcal{P}]$, its expectation or mean value called mean functional Φ_M following [16], by:

$$E\{\Phi(\varphi)\} = \Phi_M = \int \Phi(\varphi) \, d\mathcal{P}. \tag{1.140}$$

If Φ_M exists and Φ is a random field, then Φ_M will be linear but not necessarily continuous in $\mathcal{U}$. Furthermore, one can define a covariance functional Φ_C such that:

$$\Phi_C: \mathcal{U} \times \mathcal{U} \to \mathbb{C}$$

where
$$\Phi_C(\varphi, \psi) = E\{[\Phi(\varphi) - \Phi_M(\varphi)][\Phi(\psi) - \Phi_M(\psi)^*]\}. \tag{1.141}$$

It is to be noted that $\Phi_C(\varphi, \psi)$ need not exist for all $\varphi, \psi \in \mathcal{U}$, but if it does and Φ is a random field on $\mathcal{U}$, Φ_C will be a positive semi-definite bilinear form on $\mathcal{U}$, which may not be bounded. Similarly, the variance functional Φ_V, i.e. $\Phi_V: \mathcal{U} \to \mathbb{R}^+$ is defined by:

$$\Phi_V(\varphi) = \Phi_C(\varphi, \varphi) \tag{1.142}$$

and the correlation functional $\Phi_B: \mathcal{U} \times \mathcal{U} \to \mathbb{C}$ by

$$\Phi_B(\varphi, \psi) = \Phi_C(\varphi, \psi) + \Phi_M(\varphi)\Phi_M(\psi)^* = E\{\Phi(\varphi)\Phi(\psi)^*\}. \tag{1.143}$$

The random functional Φ is of second order, if $\Phi_B(\varphi, \varphi) < \infty$ for all $\varphi \in \mathcal{U}$ and if Φ is bounded there exists a Φ_B such that

$$\Phi_B(\varphi, \varphi)^{\frac{1}{2}} \leq \Phi_B \|\varphi\| \quad \text{for all} \quad \varphi \in \mathcal{U}. \tag{1.144}$$

A bounded random functional is of second order, but the converse may not hold. Corresponding to the definition of the characteristic function of a random variable (section 1.4), the characteristic random functional L_Φ will be defined by:

$$L_{\Phi(\varphi)} = E\{e^{i\Phi(\varphi)}\} = \int e^{i\Phi(\varphi)} \, d\mathcal{P} \tag{1.145}$$

and will exist for all $\varphi \in \mathcal{U}$. It is continuous, if Φ is continuous. Recalling that if x is a random variable in $[\mathcal{X}, \mathcal{F}, \mathcal{P}]$, its characteristic function $\chi_x: \mathbb{R} \to \mathbb{C}$ is given by:

$$\chi_x(t) = E\{e^{itx}\} = \int e^{itx(\omega)} \, d\mathcal{P}(\omega) \tag{1.146}$$

and for a sequence of random variables $x_1, \ldots, x_n$, the joint characteristic function $\chi_{x_1, \ldots, x_n}$ (see Doob [33] and Loève [22]) can be expressed by:

$$\chi_{x_1, \ldots, x_n}(t_1, \ldots, t_n) = E\left\{\exp\left(i \sum_{k=1}^n t_k x_k\right)\right\}. \tag{1.147}$$

44   Mathematical Preliminaries

Since the joint distribution function of the sequence $x_1, \ldots, x_n$ is uniquely determined by $\chi_{x_1,\ldots,x_n}$, it follows that if Φ is a random field, L_Φ will determine Φ uniquely to within an equivalence. Thus for a random field Φ, L_Φ is a continuous positive definite functional with $L_\Phi(0) = 1$. It can be shown that for a second-order random field, the functional is proper, iff Φ_C is an inner product on $\mathscr{U}$ (dim $\mathscr{U} \geqslant 2$). If Φ is proper and uncorrelated, it is bounded. For example, if $\mathscr{U}$ is considered as a Hilbert space and Φ is bounded, there exists a unique vector $\varphi_\Phi \in \mathscr{U}$ and two unique operators ^1T and ^2T on $\mathscr{U}$ such that:

$$\left.\begin{aligned}&(1)\ \Phi_M(\varphi) = \langle \varphi, \varphi_\Phi \rangle, \\ &(2)\ \Phi_C(\varphi, \psi) = \langle {}^1T_\Phi \varphi, \psi \rangle \\ &\text{and} \\ &(3)\ \Phi_B(\varphi, \psi) = \langle {}^2T_\Phi \varphi, \psi \rangle \quad \text{for all} \quad \varphi, \psi \in \mathscr{U}.\end{aligned}\right\} \quad (1.148)$$

These quantities are referred to as mean vector, covariance operator and correlation operator for the functional Φ, respectively (for proof see, for instance, Gudder [16]).

A certain group of random fields is sometimes important, which occurs when Φ has strongly independent values, if $\langle \varphi, \psi \rangle = 0$, which implies that:

$$E\{\Phi(\varphi)\Phi(\psi)^*\} = E\{\Phi(\varphi)\}E\{\Phi(\psi)^*\}. \quad (1.149)$$

Hence, $\Phi(\varphi)$ and $\Phi(\psi)$ are stochastically independent.

A Gaussian random field is defined by its characteristic functional analogously to the relation (1.137) given in section 1.4, i.e.:

$$L_\Phi = \exp[i\Phi_M(\varphi) - \tfrac{1}{2}\Phi_V(\varphi)]. \quad (1.150)$$

It can be shown that such a field can always be constructed for a given mean and covariance functional (see Gel'fand and Vilenkin [37], A. M. Yaglom [30] and Y. A. Rozanov [38]). The random field in which the covariance functional is of the form:

$$\Phi_C(\varphi, \psi) = r\langle \varphi, \psi \rangle; \quad r > 0 \quad (1.151)$$

is called an isonormal random field with parameter r. A unit random field Φ_u is an isonormal field with mean zero and parameter one. Such fields become significant in establishing the relationship between random fields and a field theory in quantum mechanics. Certain classes of random fields in the n-dimensional space $\mathbb{R}^n$ generated by stationary random processes were considered by Yaglom [30]. In the terminology of Yaglom complex random functions $\varphi(\mathbf{x}) = \varphi(x_1, \ldots, x_n)$ of n real variables, i.e. of a point in $\mathbb{R}^n$ form a random field in $\mathbb{R}^n$. Together with the ordinary field, a generalized random field in the sense of Gel'fand [37] and Ito [39], is then a random linear functional $\Phi(\varphi)$ with respect to φ where $\varphi = \varphi(\mathbf{x})$ is an arbitrary function of infinitely differentiable complex functions of n variables and where each φ vanishes outside a compact set. In particular, if $S(\mathbb{R}^n)$ designates a Schwartz space and if $\Phi(\varphi)$ satisfies the condition that:

$$\lim_{\varphi_1 \to \varphi} E\{|\Phi(\varphi_1) - \Phi(\varphi)|^2\} = 0. \tag{1.152}$$

Φ becomes then a generalized random field. The limit in (1.152) is to be understood in the sense of the Schwartz topology [40]. This means that in the space S in the limit,

$$(\lim_{n \to \infty} \varphi_1(n) = \varphi)$$

all functions $\varphi_1^{(n)}$ and φ vanish outside a compact set and $n \to \infty$ all partial derivatives of $\varphi_1^{(n)} \to$ the partial derivatives of the function φ on this set. Thus the space $S(\mathbb{R}^n)$ can be associated with the space of rapidly decreasing functions and if it is complete, it will form a Fréchet space. This space becomes useful in certain applications of probabilistic mechanics, since $S(\mathbb{R}^n)$ is large enough to be complete in its topology, but small enough to be closed under differentiation and multiplication by polynomials for instance. The generalized random field: in the sense of Yaglom is homogeneous, if its mean functional $\Phi_M(\varphi)$ and covariance functional $\Phi_C(\varphi, \varphi)$ are invariant under a shift transformation on S (see also [30]). Using a shift operator which shifts the argument of φ on S by a vector $\mathbf{y}$ for example, then $\tau\varphi(\mathbf{x}) = \varphi(\mathbf{x} + \mathbf{y})$ and it can be stated that a homogeneous field is characterized by:

(i) $\Phi_M(\varphi) = \Phi_M(\tau\varphi),$
(ii) $\Phi_C(\varphi_1, \varphi_2) = \Phi_C(\tau\varphi_1, \tau\varphi_2).$ $\qquad$ (1.153)

The above described generalization of stationary random processes to random fields in terms of linear functionals and particularly to homogeneous ones has been used in the statistical theory of turbulence (see of continuous media, for instance, Kolmogorov [41], Batchelor [43] and others). In the present analysis, however, it is of greater interest to define Markov random fields and particularly their equivalence with Gibbsian random fields. This will be discussed in Chapters 2 and 5 of this volume.

2
Markov Processes and Markov Random Fields

2.1 INTRODUCTION

In view of the following analysis of the behaviour of discrete media on the basis of probabilistic concepts another class of stochastic processes known as Markov processes is of utmost importance. By introducing conditional probability functions, a discrete random process $x(t)$ is said to be Markovian, if:

$$\left. \begin{array}{l} P\{\xi_n,t_n|\xi_{n-1},t_{n-1};\ldots;\xi_2,t_2;\xi_1,t_1\} \\ = P\{\xi_n,t_n|\xi_{n-1},t_{n-1}\};\quad t_n > t_{n-1} > \ldots > t_2 > t_1 \end{array} \right\} \quad (2.1)$$

indicating that the probability for the process $x(t)$ to have values ξ_n at times $t = t_n$ under the condition that its value at some earlier times are known, depends only on the most recent value $x(t) = \xi_{n-1}$ at $t = t_{n-1}$. The conditional probability function is called transition probability of the Markov process.

Thus, a discrete Markov process is completely defined by its first probability function $P(\xi,t)$ and the transition probabilities. If the initial value of the process is known, i.e. $x(0) = \xi_0$ the transition probabilities alone will describe the process. Thus the complete structure of $x(t)$ is determined by its initial state and a transition mechanism that characterizes the evolution of the process through time. The Markov property of $x(t)$ can also be expressed by conditional density functions of the form:

$$\left. \begin{array}{l} p(\xi_n,t_n|\xi_{n-1},t_{n-1};\ldots;\xi_2,t_2;\xi_1,t_1) \\ = p(\xi_n,t_n|\xi_{n-1},t_{n-1});\, t_n > t_{n-1} > \ldots > t_2 > t_1 \end{array} \right\} \quad (2.2)$$

in which the quantity on the right-hand side of (2.2) is referred to as transition probability density or briefly transition density. Denoting it by $p_{t|t'}(\xi|\xi')$, $t' < t$, $\xi, \xi' \in \mathcal{X}$ the Markov process is completely specified by these transition densities and the first-order density $p_{t_0}(\xi)$, $x \in \mathcal{X}$ for some fixed instant of time $t_0 \in T$. Although (2.2) remains valid for any parametric values of $t_1, t_2, \ldots, t_n$ so long as $t_1 < t_2 < \ldots < t_n$ the transition densities are not completely arbitrary, e.g. they must satisfy the well-known Chapman–Kolmogorov functional relation such that:

$$p_{t_3|t_1}(\xi_3|\xi_1) = \int_{-\infty}^{\infty} p_{t_3|t_2}(\xi_3|\xi_2) p_{t_2|t_1}(\xi_2|\xi_1) d\xi_2 \qquad (2.3)$$

for all $t_1 < t_2 < t_3$ and all $\xi_1, \xi_2, \xi_3 \in \mathscr{X}$.

Relation (2.3) is easily rewritten by using the definition of conditional probabilities, where:

$$p_{t_3,t_2|t_1}(\xi_3,\xi_2|\xi_1) = p_{t_3|t_2,t_1}(\xi_3|\xi_2,\xi_1) p_{t_2|t_1}(\xi_2|\xi_1)$$

and by the Markov property:

$$p_{t_3|t_2,t_1}(\xi_3|\xi_1,\xi_2) = p_{t_3|t_2}(\xi_3|\xi_2)$$

and

$$p_{t_3|t_1}(\xi_3|\xi_1) = \int_{-\infty}^{\infty} p_{t_3,t_2|t_1}(\xi_3,\xi_2|\xi_1) d\xi_2.$$

In the discrete Markov process relation (2.3) has to be expressed in terms of a sum of the transition probabilities. Although there are other random processes, which may be of importance in the description of physical phenomena, they will be discussed when required in later parts of this text.

2.2 MARKOV THEORY AND DYNAMICAL SEMI-GROUPS

Due to the importance of Markov processes in probabilistic mechanics and in the development of a general probabilistic theory of deformation and flow of discrete media, some further characteristics of such processes will be considered in this and the following sections.

In particular the concept of transition probability functions will be discussed in some detail as well as the time evolution of the material system in terms of a one-parameter semi-group of contracting linear operators acting on a Banach space. Generally such a space is generated in the description of the states of the elements of the discrete medium (see Definitions in Chapter 3). Such a semi-group is also referred to as a dynamical semi-group (see, for instance, Guz [44]). The basic results of the analytical theory of semi-groups of bounded linear operators and their application to stochastic analysis have been discussed amongst others by Yosida [7].

(i) *The Markov property and transition probabilities*

The theory of Markov processes is due to A. A. Markov [42], who first gave an analytical formulation of the well-known Brownian motion of particles in statistical fluid mechanics. This motion was characterized by a transition probability representing the probability, that a particle starting from a given point ξ is in a certain set E at time t. Subsequently, a more general theory of Markov processes was developed by A. N. Kolmogorov [2], W. Feller [28], Doob [33], and others. A comprehensive study of Markov processes is due to Dynkin [45] (see also Blumenthal and Getoor [46], Rosenblatt [24], Revuz [47]). The more advanced theory of Markov processes is concerned with the motion $x(t)$ and its path in terms of transition probability measures. The Markov processes of basic interest in probabilistic mechanics are those with a stable generating mechanism through time. Thus, if $\mathscr{X}$ is the state-space or the space of points $\omega = \{\xi_0, \xi_1, \ldots\}$ representing possible observations at fixed time instants, the possible events for which a probability is well defined will be the elements of the Borel field $\mathscr{F}$ of subsets of $\mathscr{X}$. The stable generating mechanism for a Markov process is then characterized by its transition probability function $P(\xi, E)$ which is assumed to be $\mathscr{F}$-measurable as a function of ξ for each event $E \in \mathscr{F}$ and a probability measure on the Borel field $\mathscr{F}$ for each ξ in $\mathscr{X}$. Thus, in the case of a simple Markov process, if some phenomenon is observed at time instants $t = 0, 1, 2, \ldots$ the function $P(\xi, E)$ represents the probability that an observation at time $t + 1$ falls within the set E given that at time t the observation ξ was made. Hence, the fundamental condition that relates the random process $x(t, \omega)$ in $\mathscr{X}$ or equivalently $\{x(t_i, \omega) = \xi_i;$ $t_i = 0, 1, 2, \ldots\}$ to the measure $\mathscr{P}\{x(t) \in E | x(s) = \xi\}$ is the well-known Markov principle. By using an initial probability measure $\mathscr{P}_0$ at $t = 0$ on the Borel field $\mathscr{F}$ a discrete time-parameter process can be formulated from the distribution P and a stationary transition probability function. It is apparent, that for any finite collection of event sets $E_0, E_1, \ldots, E_n \in \mathscr{F}$ the following relation will hold:

$$\begin{aligned}\mathscr{P}\{E_0 \times E_1 \times \ldots \times E_n\} &= \mathscr{P}\{\xi_0 \in E_0, \xi_1 \in E_1, \ldots, \xi_n \in E_n\} \\ &= \int_{E_0} \mathscr{P}_0 d\xi_0 \int_{E_1} P(\xi_0, d\xi_1) \ldots \int_{E_{n-1}} P(\xi_{n-2}, d\xi_{n-1}) P(\xi_{n-1}, E_n)\end{aligned} \quad (2.4)$$

where the above set functions can be extended on the basis of the Jonescu–Tulcea extension theorem (see also Doob [33]) to a measure $\mathscr{P}$ on the Borel field $\mathscr{F}_\infty$, that is generated by the product sets

$$E_0 \times E_1 \times \ldots \times E_n = \{\xi_0 \in E_0, \xi_1 \in E_1, \ldots, \xi_n \in E_n\};$$

$E_i \in \mathscr{F}$ on the space $\mathscr{X}_\infty$ of points $(\omega = \{\xi_0, \xi_1, \ldots, \xi_n\})$. Such a measure represents then the probability of observing the paths $(\omega = \xi_0, \xi_1, \ldots, \xi_n)$ of the elements of the discrete medium with time. This theorem is rather general in the sense that it does not require topological conditions to be

2.2 Markov theory and dynamical semi-groups

satisfied by the finite dimensional set functions given in (2.4). However, it restricts the Markov process to a discrete time-parameter process. In this context, it should be mentioned that another extension theorem due to Kolmogorov (see, for instance, Rényi [17], Rosenblatt [24], Dynkin [45]), which is based on measure theoretical and topological considerations and which is equally applicable to Markovian and non-Markovian processes becomes significant in probabilistic mechanics. Moreover, it also applies to both discrete and time-continuous Markov processes.

Evidently the probability density function in the manner described above, represents the transition of a material system from one state to another in one-step only. However, higher step transition probability functions, designated by $P^{(n)}$, can be generated by a recursive procedure such that:

$$P^{(n+1)}(\xi, E) = \int_{\mathcal{X}} P(\xi, d\eta) P^{(n)}(\eta, E); \quad (n = 1, 2, \ldots) \quad (2.5)$$

so that the following relation is obtained:

$$P^{(n+m)}(\xi, E) = \int_{\mathcal{X}} P^{(n)}(\xi, d\eta) P^{(m)}(\eta, E); \quad (n, m = 1, 2, \ldots). \quad (2.6)$$

The above equation represents again the Chapman–Kolmogorov functional relation. This relation is important in probabilistic mechanics, since it connects the stochastic theory with the functional analytic formulation and will be further discussed in the following chapters. So far as the stable transition mechanism of the process is concerned, one may consider the transition after a certain unit of time has elapsed from a point $\xi \in \mathcal{X}$ in more than one step into a set $E \in \mathcal{F}$ ($\mathcal{X}$ being the probabilistic function space) so that the transition probability $P^{(n)}(\xi, E)$ will be given by:

$$\left. \begin{array}{l} P^{(n)}(\xi, E) = \int_{\mathcal{X}} P^{(n-1)}(\xi, d\eta) P^{(1)}(\eta, E); \; (n = 1, 2, \ldots), \\ P^{(1)}(\xi, E) \equiv P(\xi, E) \end{array} \right\} \quad (2.7)$$

where $P^{(n)}(\xi, E) \geqslant 0$ and $P^{(n)}(\xi, \mathcal{X}) = 1$.

If there exists a non-negative set function $f(E)$ such that:

$$\left. \begin{array}{l} \text{(i)} \; f(E) \text{ is completely additive for } E \in \mathcal{F} \; \text{ and } \; f(\mathcal{X}) = 1, \\ \text{(ii)} \; \mathcal{F} \text{ is completely additive,} \\ \text{(iii)} \; \int f(d\xi) P(\xi, E) = f(E) \quad \text{for any } E \in \mathcal{F}, \end{array} \right\} \quad (2.8)$$

it is readily seen from (2.8, (iii)) that $f(E) = \int f(d\xi) P^{(n)}(\xi, E)$ is a stable distribution for any n-instants of time. It further includes the case of a deterministic transition in the sense of the ergodic theorems of Birkhoff [48]

50 Markov Processes and Markov Random Fields

and v. Neumann [49]. For example, if T denotes the one-to-one transformation of $\mathcal{X}$ onto $\mathcal{X}$, which maps $E \in \mathcal{F}$ on the set $T \cdot E \in \mathcal{F}$ in a measure preserving manner, then $f(E) = f(T \cdot E)$ will also be a stable distribution. This is readily seen by taking the characteristic function φ (section 1.3.2) such that $P(\xi, E) = \varphi(T \cdot \xi)$. The probability function $P(\xi, E)$ then conforms with that of a Markov process with a stable distribution.

Although many types of Markov processes are encountered in dealing with the behaviour of discrete materials, the processes can be approximately grouped into those that have a non-denumerable number of states and those which have a denumerable number of states. In particular the latter group commonly referred to as Markov chains are frequently employed in the probabilistic mechanics of solids and fluids. A comprehensive study of this group of processes is given by Dynkin [45], Rosenblatt [24], Revuz [47], and others. In this context, considering a stochastic process $x(t)$ assuming that at any time $t \in \mathbb{R}^+$ the transition from one state to another state depends not only on t but also on the random behaviour of a structural element of the medium, and if at time s the system is in a state i going to a state j at a later time, the probability depending on the behaviour of the system until time s is designated by p_{ij}. The Markov chain can be characterized therefore by:

$$p_{ij}(s,t) = P\{x_t = j | x_s = i\}; \quad i,j = 1, 2, \ldots \quad t, s \in T \subset \mathbb{R}^+. \quad (2.9)$$

It is to be noted, that the most important feature of a Markov chain is its reference to the discreteness of a set of time parameters since a denumerable state space can be regarded as a special case of a more general state space. However, discrete times are not a particular case of continuous time (see definition of a Markov chain by Revuz [47]). In this sense the term Markov chain refers strictly to a process, which is discrete in time, whilst a Markov process as such is one in which the time parameters are continuous. If the behaviour of the Markov chain $x(t)$ is referred to an initial time instant t_0, the corresponding initial probability distribution of some field quantity, which may be experimentally accessible, can be expressed by:

$$\overset{\circ}{p}_i = P\{x(t_0) = i\}; (i = 1, 2, \ldots) \quad (2.10)$$

so that
$$\left. \begin{array}{l} P\{x(t_0) = i, x(t_1) = i_1, \ldots, x(t_n) = i_n\} \\ = \overset{\circ}{p}_i p_{ii_1}(t_0, t_1) \ldots p_{i_{n-1} i_n}(t_{n-1}, t_n) \end{array} \right\} \quad (2.11)$$

for any $i, i_1, \ldots, i_n$ and $t_0 \leqslant t_1 \leqslant \ldots \leqslant t_n$. The Markov chain is called homogeneous, if the transition probability $p_{ij}(s,t)$ depends only on the time difference, i.e.:

$$p_{ij}(s,t) = p_{ij}(t-s); \quad (i,j = 1,2,\ldots), \quad s < t, s, t \in T \subset \mathbb{R}^+. \quad (2.12)$$

It is often convenient in the application of Markov chains to use the notion of a Markov time τ or a random time which does not depend on the future. For an

2.2 Markov theory and dynamical semi-groups

arbitrary time t the event $E\{\tau > t\}$ will be determined by the behaviour of the system until t and τ is then a Markov time. If E is an arbitrary event, the realization of which depends entirely on the behaviour of the system after τ then the behaviour is known until the time τ and the probability of E to occur coincides with the conditional probability that only the state of the system at τ is known, i.e.

$$P\{E|\xi(s), s \leq \tau\} = P\{E|\xi(\tau)\}. \tag{2.13}$$

In particular, for any $i_1, \ldots, i_n$ and $t_1 \leq t_2 \leq \ldots \leq t_n$:

$$\left.\begin{array}{r}P\{x(t_1 + \tau) = i_1, \ldots, x(t_n + \tau) = i_n | x(s), s \leq \tau\} \\ = p(\tau)p_{x_{(\tau)}, i_1}(\tau, t + \tau) \ldots p_{i_{n-1}, i_n}(t_{n-1} + \tau, t_n + \tau).\end{array}\right\} \tag{2.14}$$

Considering now the sequence $\tau_0 \leq \tau_1 \leq \ldots \leq \tau_n$ in such a manner that τ_n corresponds to states, where $x(\tau_n) = i_n$ is known. For instance, if τ_0 is the time instant when the medium is for the first time in the state i then relation (2.14) indicates that τ_1 is the instant of time when it returns to this state, τ_2 the instant for its second return, and so on. Thus, for arbitrary events $E_1, E_2, \ldots, E_n$ such that every event is completely determined by the system with corresponding time intervals $\tau_{n-1} \to \tau_n$ one can say that these are completely mutually independent events.

In this context some remarks on the previously discussed Chapman–Kolmogorov relations (2.6, 2.7) should be made. Thus by using the notation introduced in (2.9, 2.11), it is apparent that for the Markov chain one has:

$$p_{ij}(s, t) = \sum_k p_{ik}(s, u)p_{kj}(u, t); \quad (i, j = 1, 2, \ldots), \quad s \leq u \leq t. \tag{2.15}$$

If the process $x(t)$ is a homogeneous Markov chain and the time instants $t = nh$ ($n = 0, 1, 2, \ldots$ and $h \geq 0$), the probabilities $p_{ij}(nh)$ of a transition in n-steps are uniquely determined by the probabilities $p_{ij} = p_{ij}(h)$ of a one-step transition in such a manner that:

$$p_{ij}(nh) = \sum_k p_{ik}p_{kj}[(n-1)h] = \sum_k p_{ik}[(n-1)h]p_{kj}. \tag{2.16}$$

If the time parameter t is continuous and denoting $p_{ij}(0)$ specifically as:

$$p_{ij}(0) = \lim_{h \to 0} p_{ij}(h) = \begin{cases} 1 & \text{for } j = i, \\ 0 & \text{for } j \neq i, \end{cases} \tag{2.17}$$

then by assuming that the Markov chain has such a property, the probabilities will be continuously differentiable for $t > 0$. Moreover, if the following limit exists:

$$\lim_{h \to 0} \frac{p_{ij}(h) - p_{ij}(0)}{h} = q_{ij}, \quad (i, j = 1, 2, \ldots) \tag{2.18}$$

where $0 \leq q_{ij} < \infty$ for $i \neq j$ and $q_{ij} = q_{ii} = -q_i$ for $i = j$, the coefficients q_{ij} are referred to as transition density from states i to j. In terms of the Markov time τ, denoting by τ_i the instant of time when $x(t)$ leaves the state i for the first time and by τ_{ij} when $x(t)$ reaches the state j for the first time, it is evident that:

$$\tau_i = \sup_{x^{(i)}(t) = i} t\,; \quad \tau_{ij} = \inf_{x^{(i)}(t) = j} t \qquad (2.19)$$

in which $x^{(i)}(t)$ designates the process when $x(0) = i$. Hence the probability that the Markov chain $x(t)$ when leaving the state i and first moves to the state j will be given by:

$$P(\tau_{ij} = \tau_i | x(0) = i) = P\{x^{(i)}(t) = j\} = \frac{q_{ij}}{q_i}; \quad i \neq j \qquad (2.20)$$

since $x^{(i)}(t)$ is a step-function. Assuming that the above transition densities always satisfy the inequality

$$\sum_{i \neq j} q_{ij} \leq q_i, \quad (i = 1, 2, \ldots),$$

the transition probabilities $p_{ij}(t)$ under these conditions will satisfy the backward Kolmogorov equations (see Chapter 5, Doob [33], and others). Under certain restrictions such as boundedness of the q_{ij}, for instance, the forward Kolmogorov relations (Chapter 5) will also apply, which can be written as:

$$p'_{ij}(t) = \sum_k q_{kj} p_{ik}(t); \quad (i, j = 1, 2, \ldots) \qquad (2.21)$$

for the homogeneous chain and when t is continuous, $q_i \to \infty$. Hence when reaching the state i, the system leaves it instantaneously with probability one, or:

$$P\{\tau_i = 0 | x(0) = i\} = 1. \qquad (2.22)$$

For any arbitrary small time interval Δt after $t = 0$ and denoting by Δi the time spent in the state i during Δt one can write that:

$$P\left\{\lim_{\Delta t \to 0} \frac{\Delta i}{\Delta t} = 1 \Big| x(0) = i \right\} = 1 \qquad (2.23)$$

so that the state i will be a stable one, if $q_i < \infty$. Hence, the Markov chain is stable, if with probability one, the system passes, during an arbitrary finite period, a finite number of times from one state to another. The probability of passing from one state to another in one-step can also be designated by a square matrix $\underline{P}_i$ of numbers $p_{jk,i}$ characterizing the probability that at time t_{i+1}, $x(t_{i+1}) = \xi_k$ under the condition that $x(t_i) = \xi_j$. Evidently for a homogeneous chain $\underline{P}_i \equiv \underline{P}$ and since the absolute distribution at the time

2.2 Markov theory and dynamical semi-groups

$t = t_{i+1}$ is given by the summation over all possible transitions one can write:

$$q_k(t_{i+1}) = \sum_j p_{jk} q_j(t_i) \tag{2.24}$$

which in matrix form becomes:

$$\underline{q}(t_{i+1}) = \underline{P}^T \underline{q}(t_i) \tag{2.25}$$

indicating that the volume matrix of absolute probabilities that $x(t_{i+1}) = \xi_k$ is equal to the product of the transposed matrix of transition probabilities and the column matrix of absolute probabilities that $x(t_i) = \xi_j$. In this manner, one can employ a transition operator, which upon a sequential procedure yields:

$$\underline{q}(t_n) = (\underline{P}^T)^{n-1} \underline{q}(t_i) = (\underline{P}^{n-1})^T \underline{q}(t_i) \tag{2.26}$$

Denoting by $x(t_{i-n}) = \xi_k$ provided that $x(t_r) = \xi_j$ and the transition probability by $p_{jk}^{(n)}$ its corresponding matrix by $\underline{P}^{(n)}$, one obtains for the time instants $t_r \to t = t_{r+m}$ and $t = t_{r+m+n}$ the relation:

$$\underline{P}^{(m+n)} = \underline{P}^{m+n} = \underline{P}^m \underline{P}^n = \underline{P}^{(m)} \underline{P}^{(n)} \tag{2.27}$$

since $\underline{P}^{(n)} = \underline{P}^n$. Expressing this relation in scalar form yields the same relation as before (2.15), i.e. the Chapman–Kolmogorov equation in the form of:

$$p_{jk}^{(m+n)} = \sum_v p_{jv}^{(m)} p_{vk}^{(n)} \tag{2.28}$$

which can be readily extended to a non-homogeneous Markov chain. Analytical models on the basis of various types of Markov processes in the representation of the behaviour of discrete media will be considered in subsequent chapters.

(ii) *Dynamical semi-groups*

As mentioned earlier, one of the main aims of probabilistic mechanics is the formulation of the evolution through time of the relevant field quantities during deformation and flow of discrete media. It will be shown in Chapter 3 that after identifying a characteristic vector-valued field variable $z \in Z$, for instance, it is convenient to use the subspace $Z \subset \mathcal{X}$ of the general probabilistic function space to represent the motion of a structured medium.

It is assumed that the σ-algebra $\mathcal{F}^z$ can be identified in Z and that an appropriate measure $\mathcal{P}^z$ can be established. Considering such a function space $[Z, \mathcal{F}^z, \mathcal{P}^z]$ for any particular time t, the entire process will be represented by a set of function spaces leading to the notion of a product space. In particular, if $\mathbb{R}^+ = [0, \infty]$ where for each $t_r \in \mathbb{R}^+, r = 1, \ldots, N$, there corresponds a triplet $[Z, \mathcal{F}^z, \mathcal{P}^z]$, an N-fold product of such spaces forms a product space in which $z(\mathbf{X}, t)$ becomes a measurable function. For mathematical convenience this

54 Markov Processes and Markov Random Fields

space can be extended by designating it as Z_∞, $\mathscr{F}_\infty^z$, $\mathscr{P}_\infty$ so that $\mathbf{z}(\mathbf{X}, t)$ may be regarded as a time-continuous random function. However, it is of greater interest here to consider an alternative formulation in terms of a one-parameter family of transformations L_t such that:

$$L_t: Z \to Z \quad \text{for} \quad \forall t \in \mathbb{R}^+ = [0, \infty]; \quad \mathbf{z}_t(\mathbf{X}) \in Z. \tag{2.29}$$

In this sense a deformation process, for instance, for a given solid can be defined by the automorphism L_t for all $t \in \mathbb{R}^+$. However, a deformation process in general will be formed by a family of L_t and hence an endomorphism of transformations.

Analogously, if Z is considered to be a velocity space $\mathscr{V}$ in the analysis of the flow field of a given discrete fluid, the velocity field $\mathbf{z} = \mathbf{v}(\mathbf{r}, t)$ can be described in terms of the family $\{L_t\}$ with each L_t representing a unique history of the possible motion. The σ-algebra $\mathscr{F}_\infty$ of the product space mentioned above has a countably finite number of Borel sets $E_1(\mathbf{z}) \times E_2(\mathbf{z}) \times \ldots \times E_N(\mathbf{z})$ corresponding to the sequence $t_1, t_2, \ldots, t_N \in \mathbb{R}^+$ and where each Borel set is obtained from the other by L_t or its inverse. If the sequence $t_1 < t_2 < \ldots < t_N$, it can be stated that:

$$L_{\Delta t_r} E_r(\mathbf{z}) = E_{r+1}(\mathbf{z}); \quad r = 1, 2, \ldots, N-1, \quad \Delta t_r = t_{r+1} - t_r. \tag{2.30}$$

It can be easily shown that, if $\mathscr{P}^z$ is a regular measure on E_r and L_t is defined so as to satisfy (2.30), E_{r+1} is also $\mathscr{P}^z$-regular measurable. This statement can be generalized to any Borel sets $E_r(\mathbf{z}); r = 1, \ldots, N$ in $\mathscr{F}^z$. Hence, it may be concluded that at any time during a deformation process, the random deformations are $\mathscr{P}^z$-regular measurable. In view of the application of the Markov theory to such a deformation process, it is more fundamental to introduce the notion of conditional probabilities. Hence, to each automorphism $L_{\Delta t_r}$ there corresponds a conditional probability measure $\mathscr{P}\{E_{r+1}|E_r\}$ such that whenever

$$L_t E_r(\mathbf{z}) = E_{r+1}(\mathbf{z}) \tag{2.31}$$

holds, then:

$$\mathscr{P}^z\{E_{r+1}\} = \mathscr{P}^z\{E_{r+1}|E_r\} \mathscr{P}^z\{E_r\}. \tag{2.32}$$

The latter equation is readily generalized to:

$$\mathscr{P}^z\{E_n\} = \mathscr{P}^z\{E_1\} \prod_{r=1}^{n-1} \mathscr{P}^z\{E_{r+1}|E_r\}. \tag{2.33}$$

This relation is valid for any sequence $E_1 \supset E_2 \supset \ldots \supset E_{N-1} \supset E_N$ corresponding to the time sequence $t_1 < t_2 < \ldots < t_r < \ldots t_N$ and a set of conditional probabilities $\mathscr{P}\{E_{r+1}|E_r\}, r = 1, 2, \ldots, N-1$. The above relations (2.32, 2.33) are of utmost significance in the probabilistic mechanics theory of discrete media since they permit the determination of the probability distribution of certain field variables at any time during the process

2.2 Markov theory and dynamical semi-groups

(deformation or flow), if the distribution at any other time or the initial one is known. This is also recognized by rewriting (2.32) more explicitly in the following manner:

$$\mathscr{P}^z\{E_{r+1}, t_{r+1}\} = \mathscr{P}\{E_{r+1}, t_{r+1} | E_r, t_r\} \mathscr{P}^z\{E_r, t_r\} \tag{2.34}$$

in which E_r corresponds to $t_r \in \mathbb{R}^+$ and E_{r+1} to $t_{r+1} > t_r$. Denoting the space of all measures $\mathscr{P}^z\{E_n\}$ by $L(0,1)$, it is seen that this conditional probability on $L(0,1)$ to $L(0,1)$ is a contraction, i.e. $\mathscr{P}\{\cdot|\cdot\} \leq 1$. In the theory of Markov processes the probability transition functions can also be introduced in the following manner. The function $P(s, \zeta, t, E)$ with probability one is a transition function, if

$$P(s, \zeta, t, E) = \mathscr{P}\{z(t) \in E | z(s) = \zeta\} \tag{2.35}$$

satisfying the following conditions:

(i) $P(s, \zeta, s, E) = \begin{cases} 1 & \text{for } \zeta \in E \\ 0 & \text{for } \zeta \notin E; \end{cases}$

(ii) $P(s, \zeta, t, E) \leq 1$;

(iii) for fixed s, t and $\zeta \in Z$, $P(s, \zeta, t, E)$ is a probability measure on Z;

(iv) for fixed s, t and $E \in \mathscr{F}^z$, $P(s, \zeta, t, E)$ is a Z-measurable function of $\zeta \in Z$;

(v) $P(s, \zeta, t, E) = \int_Z P(s, \zeta, \tau, d\eta) P(\tau, \eta, t, E); \quad s \leq \tau \leq t.$

$\tag{2.36}$

This relation is the Chapman–Kolmogorov equation. It is important to note that such a transition function will always exist, if the subspace Z of $\mathscr{X}$ is separable and the probability distributions

$$\mathscr{P}_t\{E\} = \mathscr{P}\{z(t) \in E\}, E \in \mathscr{F}$$

are perfect measures. This topology in Z induced by the bounded events is a weak topology (see also Guz [44]).

If the Markov process under consideration is time-homogeneous as, for example, in a steady-state deformation of a structured solid or the uniform flow of a discrete fluid, the transition function reduces to:

$$P(t, \zeta, E) \equiv P(0, \zeta, t, E) \tag{2.37}$$

so that Chapman–Kolmogorov equation takes the form of:

$$P(t+s, \zeta, E) = \int_Z P(t, \zeta, d\eta) P(s, \eta, E). \tag{2.38}$$

If the space $Z \subset \mathscr{X}$ is considered to be discrete, one obtains accordingly the matrix equation:

$$\underline{P}(t+s) = \underline{P}(t)\underline{P}(s) \tag{2.39}$$

indicating that a temporally homogeneous process exhibits the semi-group property. Using an operational formalism and introducing a transition operator T parametrized by the time t one can write that:

$$T_t[f(\zeta)] = \int_Z f(\eta) P(t, \zeta, d\eta), \quad f \in C(Z) \tag{2.40}$$

where $C(Z)$ is a Banach space of all bounded continuous functions $f(\zeta)$ on Z with a norm

$$\|f\| = \sup_{\zeta \in Z} |f(\zeta)|.$$

It follows from the properties (ii) and (v) above that $\{T_t, t \geq 0\}$ is a contraction semi-group of operators on $C(Z)$, i.e.:

$$\left.\begin{array}{ll} \text{(i)} & T_{t+s}[f(\zeta)] = T_t T_s[f(\zeta)], \\ \text{(ii)} & T_0 = I \quad \text{(Identity operator)}, \\ \text{(iii)} & \|T_t\| \leq I. \end{array}\right\} \tag{2.41}$$

This semi-group is often referred to as a dynamical semi-group (see, for instance, Kossakowski [50], Mackey [51] and Guz [44]).

It is noted from the properties of the T_t-operator that the condition of the transition function $P(t, \zeta, E)$ to be stochastically continuous is equivalent to the condition that the semi-group of operators T_t be continuous, i.e.:

$$s\text{-lim}_{t \to t_0} T_t[f(\zeta)] = T_{t_0}[f(\zeta)]; \quad f(\zeta) \in C(Z). \tag{2.42}$$

It is to be noted that in the context of Markov theory, one can use the weak limit corresponding to the weak topology of $C(Z)$ since in this case these two limits are equivalent (see also Butzer and Berens [52]). An important concept in the theory of semi-groups in a Banach space, which should be mentioned is the infinitesimal generator A for an operator $T_t: \mathscr{U} \to \mathscr{U}$, i.e.:

$$Au = s\text{-lim}_{t \downarrow 0} t^{-1}(T_t - I)u.$$

A theorem by Yosida [7] concerning contracting C_0-class semi-groups is of further interest here. It can be stated as follows:

"if $\mathscr{U}$ is a Banach space and $\{T_t\}$ a one-parameter contracting C_0-class semi-group of linear operators acting in $\mathscr{U}$, the infinitesimal generator A of this semi-group will have the following properties":

(i) A is a linear operator in $\mathscr{U}$ with a domain:

$$\mathscr{D}(A) = \{u \in \mathscr{U}: s\text{-}\lim t^{-1}(T_t - I)u \text{ exists in } \mathscr{U}\}.$$

(ii) Thus, if $u \in \mathscr{D}(A)$, $t \mapsto T_t u$ is a strongly differentiable function of the variable $t \geq 0$ such that

$$\frac{d}{dt}(T_t u) = AT_t u = T_t Au, \quad t \geq 0$$

then the solution is the operator function $T_t = e^{At}$.

2.3 MARKOV RANDOM FIELDS

The generalization of stationary random processes to random fields and in particular to homogeneous ones, has been briefly discussed in Chapter 1. In the present analysis it is of greater interest, however, to consider Markov random fields and in particular their equivalence to Gibbsian random fields. A detailed discussion on these equivalent fields will be given in Chapter 5. For the present purpose the generalization of a stochastic function $x(t, \omega)$ or $x_t(\omega)$, $t \in T$ to a Markov random field can be achieved by using the fundamental concepts of conditional probabilities and their corresponding distributions. This approach has been taken by Dobrushin [53], Averintsev [54], Rozanov [55].

Thus, following Dobrushin, consider X_t to be a random field, i.e. a random function parametrized by $t \in T^r$ taking values ξ_t in a finite set X. T^r is considered to be a finite dimensional lattice on which a domain D as a subset of T^r is regarded to represent a bounded physical domain. Denoting by $D = \{t_1, \ldots, t_n\} \subset T^r$, a finite subset of T^r, the corresponding distribution of probabilities will be of the form:

$$P\{x_{t_1}(\omega) = \xi_{t_1}, \ldots, x_{t_n}(\omega) = \xi_{t_n}\} = P_D(\xi_{t_1}, \ldots, \xi_{t_n}). \tag{2.43}$$

These distributions $P_D(\cdot|\cdot)$ form a system, corresponding to all finite subsets $D \subset T^r$ satisfying the conditions of consistency, i.e. that for any $D' = (t_1, \ldots, t_m) \subset D$:

$$\sum_{\xi_{t_{m+1}} \cdots \xi_{t_n} \in X} P_D(\xi_{t_1}, \ldots, \xi_{t_n}) = P_{D'}(\xi_{t_1}, \ldots, \xi_{t_m}). \tag{2.44}$$

The distribution $P_D(\cdot, \cdot)$ in (2.43) is called by Dobrushin the distribution of the random field. Similarly, as in Chapter 1, a homogeneous field is characterized by the invariance of the above distributions $P_D(\cdot, \cdot)$ with respect to the group of translations in T^r. This means that for any shift vector $\tau \in T^r$ one has:

$$P_D(\xi_{t_1}, \ldots, \xi_{t_n}) = P_{D+\tau}(\xi_{t_1+\tau}, \ldots, \xi_{t_n+\tau}). \tag{2.45}$$

Introducing conditional probabilities and designating the corresponding set

58 Markov Processes and Markov Random Fields

of probability distributions by $Q = \{q_{D,x(t)}(\xi_{t_1}, \ldots, \xi_{t_n}); \xi_i \in X, i = 1, \ldots, n\}$ for each subset $D \subset T^r$ and to each function $x(t) = \xi_t, t \in T^r \setminus D$ with values in X, results in:

$$\underline{P}\{x(t_1) = \xi_{t_1}, \ldots, x(t_n) = \xi_{t_n} | x(t), t \in T^r \setminus D\} = q_{D,x(t)}(\xi_{t_1}, \ldots, \xi_{t_n}). \quad (2.46)$$

Hence the system Q includes the conditional probabilities for the values of $x(t)$ on D under the condition that the values of the random field outside of this domain are known. Denoting by D_d the set of points $(T^r \setminus D)$, d being an integer, a random field as a d-Markovian field for $d > 0$ is then defined by:

$$q_{D,x(t)}(\xi_{t_1}, \ldots, \xi_{t_n}) = q_{D,x'(t)}(\xi_{t_1}, \ldots, \xi_{t_n}) \quad \text{for} \quad x(t) = x'(t), t \in D_d. \quad (2.47)$$

It can be readily shown that a complete chain of order d, for instance, is a Markov field with $r = 1$. The conditional probabilities can also be expressed in their basic forms.

It is always assumed that for any fixed set D and $(\xi_{t_1}, \ldots, \xi_{t_n})$, the variable $q_{D,x(t)}(\cdot, \cdot)$ is a measurable function of $x(t)$, if in the space of all functions $\{x(t), t \in T^r \setminus D\}$, the σ-algebra $\mathscr{F}_{T^r \setminus D}$ is used which is generated by the following sets:

and
$$\left. \begin{array}{l} \{x(t): \xi_{t_1} = \xi_1, \ldots, \xi_{n-1}; \ \xi(t_n) = \xi_n\} \\ \{t_1, \ldots, t_n\} \subset T^r \setminus D; \ \xi_1 \in X, \xi_n \in X. \end{array} \right\} \quad (2.48)$$

It should be noted that the homogeneity property of the d-Markov field can be identified in certain cases and can then be expressed, if for each vector $\bar{t} \in T^r$:

$$q_{D,x(t)}(\xi_{t_1}, \ldots, \xi_{t_n}) = q_{D+\bar{t},x(t+\bar{t})}(\xi_{t_1+\bar{t}}, \ldots, \xi_{t_n+\bar{t}}). \quad (2.49)$$

The above representation of a Markov random field is based on a lattice structure. It is apparent that for many problems in the probabilistic mechanics of discrete media a more general representation may be required. Although it is recognized that in the mathematically opposite case of the parameter space, i.e. the continuous multi-dimensional parameter space will lead to a random continuum field theory, a more general formulation should be based on the random geometry of the physical domain, i.e. as, for instance, in the case of amorphous solids, polycrystalline solids and molecular fluids.

Such an approach is then based on the notion of nearest-neighbour interactions and thus belongs to statistical physics (see Chapter 5). The description of a Markov random field in terms of the theory of probability is founded on purely mathematical concepts (see Lévy [56] and Wong [57]). For the purpose of probabilistic mechanics, however, and in particular for considerations of microdynamical phenomena (see Chapter 4) another approach is taken that permits the development of rigorous evolution relations in a multi-dimensional time space. This notion distinguishes clearly between a macro- and micro-time scale at which the actual macroscopic and microscopic phenomena take place.

3
General Formulation of Probabilistic Mechanics

3.1 INTRODUCTION

The mathematical preliminaries and the basic stochastic processes including Markov processes were briefly discussed in the preceding chapters. It is the principal aim of this chapter to give a general formulation of probabilistic mechanics leading subsequently to a more detailed description of the behaviour of discrete media during deformation and flow. Discrete media have two distinct features. First, there exists a multitude of singular surfaces (internal interfaces) within a given domain of the macroscopic material body. Second, the elements of the structure of real materials have a finite size and exhibit random physical and configurational properties that cannot readily be brought into line with the conventional deterministic macroscopic relations. Although several theories have been tried on the basis of statistical mechanics and the principles of continuum mechanics [58, 59], the ensuing analysis still refers to a network of particles (mass points) or a well-ordered cell-like structure that does not reflect the actual discreteness of the materials.

In view of these facts the probabilistic mechanics theory considers from the onset the significant geometrical and physical properties of a medium with microstructure, as random variables or functions of such variables. In developing a general formulation of probabilistic mechanics, it becomes necessary to study the evolution of the pertinent field variables with time and the associated probability measures. For convenience the notion of an abstract dynamical system representing the motion of the discrete medium in the physical space will be adopted throughout. As mentioned earlier, most of the mathematical models employed in the representation of the behaviour of discrete materials can be drawn from the theory of Markov processes. Furthermore, the collective motion of elements of the microstructure is generally of interest and correspondingly the collection of random variables and their evolution on a common probability space leads to considerations of random fields. Before stating the fundamental postulates of probabilistic mechanics and giving the axiomatic definitions of various field variables

60   General Formulation of Probabilistic Mechanics

involved in the subsequent theory, it may be instructive to discuss briefly the meaning of states and observables in the context of this theory.

3.2 STATES OF ELEMENTS OF THE STRUCTURE

Analogously to stochastic quantum mechanics two concepts of classical statistical mechanics are fundamental in the present formulation of probabilistic mechanics. These notions concern states and observables. Considering a given medium to be in a specific state or condition, it will be usually subjected to some experimental procedure in order to obtain measurements of observable quantities of the elements of its structure. For a given medium in a specific state s of the set $S = \{s, s_1, s_2, \ldots\}$, one may measure a single observable A or a set of observables $\mathcal{A} = \{A, A_1, A_2, \ldots\}$. Although the result of measuring $A \in \mathcal{A}$ may give a certain number, in general, it will be necessary to perform a number of experiments so that a statistical distribution of A is obtained. Thus given $A \in \mathcal{A}$ and the corresponding state $s \in S$ one obtains the distribution $P(A, s, E)$ which can be interpreted as the probability of the observable A having a value in the Borel set E, when the system is in the state s. This distribution is also a probability measure $\mathcal{P}$ on the space $\mathbb{R}^1$. In dealing with stochastic quantum mechanics (see Gudder [16]) the notion of an event structure can be used in which the events of a physical system are taken as the primitive axiomatic elements. The events correspond to physical phenomena, which may occur or may not occur. In particular for the type of observables which attain at most the two possible values 0 and 1, one can speak of a generalized event of an observable A having the distribution $P\{A, s: 0 < s < 1\} = 1$ for every $s \in S$.

In the present analysis, which is aimed at the representation of a physical system by employing frequently Markovian processes, it may be useful to introduce the notion of a propositional logic or briefly logic. This logic can be defined (see also Kossakowski [50], Mackey [51] and Yosida [7]) as a partially ordered set $\mathscr{E}$ with the smallest and greatest element 0 and 1, respectively and which has the following characteristics:

(i) The set $\mathscr{E}$ is orthocomplemented, i.e. there is a map $a \mapsto a'$ of $\mathscr{E}$ into itself such that $a \leqslant b$ implies $b' \geqslant a'$, $(a')' = a$ and $a \wedge a' = 0$ for all $a \in \mathscr{E}$. The symbol $\wedge$ designates inf. of a and a' or the greatest lower bound on $\mathscr{E}$.

(ii) For each sequence $a_1, a_2, \ldots$ of pairwise orthogonal propositions, there exists in $\mathscr{E}$ its least upper bound $V_{i=1}^{\infty} a_i$; two propositions $a, b \in \mathscr{E}$ are orthogonal or symbolically $a \perp b$, if $a \leqslant b'$ or equivalently $b \leqslant a'$.

(iii) $\mathscr{E}$ is orthomodular, i.e. $a \leqslant b$ implies $b = a \vee c$ (sup. of a and c) for some $c \in \mathscr{E}$, $c \perp a$.

Hence a statistical state can be defined as a non-negative real function s on $\mathscr{E}$

such that $s(1) = 1$ and

$$s(V_{i=1}^{\infty} a_i) = \sum_{i=1}^{\infty} s(a_i) \tag{3.1}$$

for each sequence of mutually orthogonal propositions. It follows from (iii) that $a \leqslant b$ always leads to $s(a) \leqslant s(b)$ and thus maps $\mathscr{E}$ into the interval $[0, 1]$. An observable A can be considered as a map from the Borel sets of the real line $\mathbb{R}^1$ into $\mathscr{E}$, i.e. $A: \mathscr{B}(\mathbb{R}^1) \to \mathscr{E}$ such that:

(i) $A(\mathbb{R}^1) = 1$.
(ii) If $E \cap F = \phi$ then $A(E) \perp A(F)$.
(iii) If $E_i \in \mathscr{B}(\mathbb{R}^1)$ is a sequence of mutually disjoint sets, then:

$$A\left(\bigcup_{i=1}^{\infty} E_i\right) = V_{i=1}^{\infty} A(E_i). \tag{3.2}$$

The number $s(A(E))$ can be regarded as the probability that a measurement of an observable A of the system in the state s will lead to a value in the Borel set E. One can express the expected value of A in the state s of the system by:

$$\langle A, s \rangle = \int_{\mathbb{R}^1} ts(A(dt)) \tag{3.3}$$

if such an integral exists. A will be bounded, if its spectrum or sp A, which is defined as the smallest closed set $E \subset \mathbb{R}^1$ such that $A(E) = 1$, is bounded. One can then define a norm for A, i.e.:

$$\|A\| = \sup\{|t|: t \in \operatorname{sp} A\}. \tag{3.4}$$

In the present formulation only bounded observable will be considered. It is to be noted, that generally in order to distinguish between the quantum and classical mechanics a specific logic in each of these theories is required. Thus in the terminology of quantum mechanics events can be thought of as propositions that correspond to true or false experiments. However, in the present formulation stressing the stochastic properties of a physical system the term event will be employed throughout. Thus in the terminology of classical mechanics propositions represent Borel subsets of the phase-space of the system. Since S is in the present analysis considered as a set of all mechanical states of the medium, the preferred term is state space. In most cases of the subsequent formulation the topological structure of S will be assumed to be of a simple form, i.e. excluding more complicated structures such as differential manifolds. In general the set S of all states of the medium may be considered as a norm-closed convex subset of a suitably ordered (partially) Banach space. The fundamental problem of the time-evolution of states from the point of

62 General Formulation of Probabilistic Mechanics

view of Markovian processes will be considered later after the definitions of the relevant field quantities have been given.

3.3 FUNDAMENTAL CONCEPTS AND DEFINITIONS

The formulation of probabilistic mechanics is based on four fundamental concepts, which are given in the form of postulates (see also refs. [36, 60]):

POSTULATE 1: Three measuring scales are used in which the smallest refers to a microelement of the structure, an intermediate one called mesodomain containing a statistical ensemble (Gibbsian) of microelements and finally a finite number of non-intersecting mesodomains forms the macroscopic material body or volume.

These scales can be used for the spatial and temporal distinction of elements within an ensemble and for the entire body.

POSTULATE 2: All field quantities pertaining to a microelement are random variables or functions of such variables.

POSTULATE 3: Stresses, strains, rates of strain, etc., and analogous quantities for fluids are generalized so that the response of a microelement includes interaction forces between contiguous elements. Such forces are in general derivable from a bond or interaction potential.

POSTULATE 4: A material operator is used that contains in its argument characteristics of the specific medium under consideration.

In solids and in fluids it connects the stresses with deformations and or the derivatives of deformations. In fluids an additional operator is required in the molecular dynamics description of such systems.

It is seen from Postulate 1 that the finiteness of a microelement of the structure is recognized and hence the analysis is to be carried out in accordance with the principles of statistical mechanics.

In order to characterize the behaviour of discrete systems in terms of an abstract dynamical system and to give the latter an appropriate structure, it is assumed that a set of admissible state vectors exists that describes the mechanical states of an ensemble of elements at any given time during deformation or flow. In view of Postulate 2 the corresponding state space is a probabilistic function space. According to Postulate 3 the quantities representing the overall deformations of a solid, and those entering in the description of flow phenomena of discrete fluids can be generalized so that interaction effects between elements can be taken into account. Postulate 4 merely stipulates the use of an operational formalism instead of the conventional constitutive relations. In accordance with these postulates the following axiomatic definitions are given:

3.3 Fundamental concepts and definitions

Def. 1: *Microelement*
$^\alpha A \in \{^\alpha A\}$: is a set of mathematical manifolds. Each member of the set represents a microelement. In the following, microelements will also be denoted simply by $\alpha, \beta, \ldots$

Since in continuum theory the notion of material points is employed, the latter can be defined in the present context as follows:

Def. 2: *Material point*
$^\alpha a \stackrel{\text{df}}{=}$ as an element of $^\alpha A \in \{^\alpha A\}$. Thus $^\alpha a$ is a mathematical representation of an element or unit that is smaller than a microelement. It may refer to a molecule or an atom depending on the requirements and scale used in the analysis.

Def. 3: *Reference frame*
$k \in \{K\}$: is a set of reference frames k, each of which is fixed within an individual microelement α or parts of it. The Euclidean space $\mathbb{R}^3$ can be related to the k-frames by time-dependent orientation transformations.

Def. 4: *Configuration*
A configuration is the image of α at time t; $\mathbf{r}(^\alpha A, t) \in \mathbb{R}^3$ or briefly $^\alpha \mathbf{r} \in \mathbb{R}^3$.

Def. 5: *Location*
A location is the image of $^\alpha a$ at time t relative to the local frame $k \in \{K\}$; $^\alpha \mathbf{r}(^\alpha a, k, t) \subset \mathbb{R}^3$. Thus configuration refers to a microelement and location to a material point.

Def. 6: *Motion*
A motion of $^\alpha A \stackrel{\text{df}}{=} \{\mathbf{r}(^\alpha A, t): -\infty < t < \infty\}$ and for $^\alpha a \stackrel{\text{df}}{=} \{\mathbf{r}(^\alpha a, k, t); -\infty < t < \infty\}$ in the discrete and time-continuous case. In probabilistic mechanics these motions are taken as stochastic processes $\{x_t\}$ defined for each microelement $^\alpha A$ or material point $^\alpha a$ as the case may be.

Def. 7: *Mesodomain*
A mesodomain M is defined as a countable set of microelements $^\alpha A$ or $M = \{^\alpha A\}$. Each member of the set of mathematical manifolds $\{M\}$ represents a mesodomain.

Def. 8: *Macrodomain*
A macrodomain $\mathcal{M}$ is the union of disjoint mesodomains and the mathematical manifold representing the macroscopic material body.

The mesodomain is considered analogous to a Gibbsian ensemble of microelements in the sense of Def. 1 or Def. 2 as required in the analysis. It is assumed that the statistics of all geometrical and physical parameters within a

64 General Formulation of Probabilistic Mechanics

mesodomain in the physical space and correspondingly in the abstract space are position independent, i.e. a set or sets of α will have properties which are statistically homogeneous.

Def. 9: *Volume*

(i) Volume of a microelement $\alpha \in \{\mathscr{A}\}$ at time t:

$$^\alpha v \equiv v(\alpha, t) = \int_{\mathbf{r}(\alpha, t)} d^3\mathbf{r},$$

$$^\alpha v_0 \equiv v(\alpha, 0) = \int_{\mathbf{R}(\alpha, 0)} d^3\mathbf{R}$$

where $\mathbf{r}, \mathbf{R}$ refer to the current and initial positions in the Euclidean frame, respectively.

An intersecting system of elements $\alpha, \beta, \ldots \in \{\mathscr{A}\}$ has the volume $v(\alpha \cap \beta) = 0$ in hard-sphere models of simple fluids, whereas in systems with strong bonding $v(\alpha \cap \beta) \neq 0$.

(ii) Volume of a mesodomain $^M v$ at time t:

$$^M v \equiv v(M, t) = \int_{\mathbf{r}(M, t)} d^3\mathbf{r},$$

$$^M v_0 \equiv v(M, 0) = \int_{\mathbf{R}(M, 0)} d^3\mathbf{R}.$$

(iii) Volume of the macrodomain at time t:

$$V \equiv {}^{\mathscr{M}}v = v(\mathscr{M}, t) = \int_{\mathbf{r}(\mathscr{M}, t)} d^3\mathbf{r}$$

and by (i) and (ii):

$$^{\mathscr{M}}v = \sum_{\{M\}} {}^M v = \sum_{\{M\}} \sum_{\{\mathscr{A}\}} {}^\alpha v$$

Def. 10: *Mass density*

(i) *Microelement*

$^\alpha \mu \in \mathscr{M}$: $^\alpha \mu$ is a scalar element of the set $\mathscr{M}$ representing the mass of the microelement $\alpha \in \{\mathscr{A}\}$ $^\alpha \rho \in \mathbb{R}$: $^\alpha \rho$ is a scalar element of the set $\mathscr{R}$ representing the mass density of $\alpha \in \{\mathscr{A}\}$ defined as $^\alpha \rho = {}^\alpha \mu / {}^\alpha v$.

3.3 Fundamental concepts and definitions 65

Equivalently using the Lebesgue–Stieltjes integral,

$$^\alpha\mu = \int_{\mathbf{r}(\alpha,t)} {}^\alpha\rho\, d^3\mathbf{r} = {}^\alpha\rho \int_{\mathbf{r}(\alpha,t)} d^3\mathbf{r}.$$

(ii) *Mesodomain*

From above, the mass of a mesodomain is given by:

$$^M\mu = \int_{\mathbf{r}(M,t)} {}^M\rho\, d^3\mathbf{r} = {}^M\rho\, {}^M v = \sum_{\{\mathscr{A}\}} {}^\alpha\rho\, {}^\alpha v$$

and hence the mean density becomes:

$$^M\rho = \int_{X^\rho} \rho\, d^M\mathscr{P}(\rho); \qquad \int_{X^\rho} d^M\mathscr{P}(\rho) = 1$$

in which the Lebesgue–Stieltjes integral extends over the subspace $X^\rho \in \mathscr{X}$ on $\mathscr{M}$, which is embedded in the general state space $\mathscr{X}$ and where $\mathscr{P}(\rho)$ is the Lebesgue measure on X^ρ. Similar definitions apply to the macrodomain. It should be noted that in the formalism of molecular dynamics, the quantities defined by Defs. 9, 10 can be interpreted in a somewhat different manner (see later discussions in Chapter 5).

Def. 11: *Function space $\mathscr{X}$ and measure $\mathscr{P}$*

(i) The usual state space s is identified with a probabilistic function space $\mathscr{X}$.

(ii) The state vector ${}^\alpha\mathbf{s} \in \mathscr{X}$ represents a set of r-parameters describing the mechanical states $i = 1, \ldots, r$ of an element α. Thus ${}^\alpha\mathbf{s}$ is an outcome or an elementary event in the probability space $\mathscr{X}$ as a result of the statistical experiment ${}^\alpha A$.

(iii) The event E is a set of state vectors within an experimental range $\Delta\mathbf{s}$ of measurements such that:

$$E = \{{}^\alpha s^i : s^i < {}^\alpha s^i < s^i + \Delta s^i; i = 1 \ldots r\}; \quad E \in \mathscr{F}$$

where $\mathscr{F}$ is the σ-algebra of open spheres or events of the set of state vectors.

(iv) $\mathscr{P}$ is the probability measure of the events E.

(v) The triple $\mathscr{B} = \mathscr{B}[\mathscr{X}, \mathscr{F}, \mathscr{P}]$ defines an abstract dynamical system, where $\mathscr{B}$ usually is a topological vector space.

It is to be noted that the components of the state vector ${}^\alpha\mathbf{s}$ in general may be scalar or higher dimensional functions. For convenience of the analysis in probabilistic mechanics one may use one or more components of the state vector $^\alpha$s. In this case attention will be given to the corresponding subspace of $S = \mathscr{X}$ and where the event structure is then identified accordingly together with an appropriate probability measure. It has to be recognized, however, that in experimental observations one can only measure quantities within a finite range Δs_n and thus the sets E become window sets. Since there is only a finite number of such sets however large it may be, it becomes necessary to index them so that:

$$E_n = \{{}^\alpha s_n^i : s_n^i < {}^\alpha s_n^i < \Delta s_n^i; n \in Z^+; \quad (i = 1 \ldots r)\}$$

giving a σ-algebra $\mathscr{F}$ that can be identified as a subalgebra of the σ-algebra $\mathscr{B}$ on $\mathscr{X} = \mathbb{R}^r$.

3.4 THE STRUCTURE OF THE PROBABILISTIC FUNCTION SPACE $\mathscr{X}$

In establishing the probabilistic structure of deformation and flow of discrete media the concept of the state space as a probabilistic function space $\mathscr{X}$ should be further clarified. This notion is well founded in classical and quantum statistical mechanics (see, for instance, Khintchine [61], Kampé de Fériet [62], Kac [63] and others). Thus, the mechanical states of a microelement $^\alpha A$ or α can be represented by an r-dimensional state vector, the components of which are real valued functions of the geometrical and thermomechanical parameters characterizing the medium. For the representation of a mesodomain a set of these state vectors $\{{}^\alpha s: {}^\alpha s_i \to {}^\alpha s^i, \alpha = 1, \ldots, N; i = 1, \ldots, r\}$ is required. The latter forms the state space $S = \mathscr{X}$ which is assumed to be locally compact. An analogous structure of the state space S can be given to that of the general state space in classical mechanics [63] by noting that the vector $^\alpha$s can only be specified within certain limits. This is due to experimental constraints and the accuracy with which relevant observations can be carried out. Hence the state of a microelement will be characterized by $s < {}^\alpha s < s + \Delta s$, where $^\alpha$s is a specific value for the element α and Δs the range of experimental observations. This indicates that by presetting this range for a particular experiment one can only state the number of microelements that will have their mechanical states within that range. One obtains therefore a subset $E_n \subset S = \mathscr{X}$, which includes the states within this range only. The subset E_n can also be considered as an open sphere so that:

$$E_n = \{\mathbf{s}_n < {}^\alpha \mathbf{s} \leqslant \mathbf{s}_n + \Delta \mathbf{s}_n\}; \bigcup_n E_n = \mathscr{X}, E_n \bigcap E_k = \phi, n \neq k \quad (3.5)$$

If $\mathscr{X}$ is locally compact, these subsets are also compact and bounded under closure. With this interpretation one can define a class of sets $\mathscr{F}$ with the following properties:

(i) $E_n \in \mathscr{F} \Rightarrow E_n' \in \mathscr{F}$; E' = complement of E,

(ii) $E_n \in \mathscr{F} \Rightarrow \bigcup_1^\infty E_n \in \mathscr{F}$,   (3.6)

(iii) $\mathscr{X} \in \mathscr{F}$

3.4 The structure of the probabilistic function space $\mathscr{X}$ 67

showing that the class $\mathscr{F}$ forms a σ-algebra as discussed in Chapter 1. The elements E_n of $\mathscr{F}$ are Borel sets and the space $\mathscr{X}$ together with $\mathscr{F}$ so defined forms a measurable space $[\mathscr{X}, \mathscr{F}]$. Since the main concern of probabilistic mechanics is the analysis of the random motion during deformation and flow of discrete media, an appropriate measure on the subsets of $\mathscr{X}$ must be chosen such that $0 \leqslant \mathscr{P}\{E_n\} \leqslant 1$; $\mathscr{P}\{E_n\} = 0$ if $E_n = \phi$ and $\mathscr{P}\{\mathscr{X}\} = 1$. In accordance with probability theory this indicates that $\mathscr{P}$ is the distribution of the relevant field quantities. By using this measure one can designate a measure space by the triple $[\mathscr{X}, \mathscr{F}, \mathscr{P}]$ which can also be interpreted as a probability space.

As mentioned earlier, since the state vector $^\alpha\mathbf{s}$ contains several components each of which belongs to a subspace of $\mathscr{X}$ it is often convenient to introduce such subspaces in the analysis. Thus, for instance, one may choose in the case of the deformations of structured solids a deformation space $\mathscr{U}$ and a corresponding stress space Σ as subspaces of $\mathscr{X}$. Similarly in the analysis of the flow of discrete fluids one can use a configuration space $\mathscr{C}$ or the velocity space $\mathscr{V}$ and correspondingly a force or stress space Σ. Consequently the state vectors in either case may be expressed in an approximate form by:

$$\text{(a)} \quad {^\alpha}\mathbf{s} = \begin{bmatrix} {^i}\sigma(t) \\ {^s}\sigma(t) \\ {^i}\mathbf{u}(t) \\ {^s}\mathbf{u}(t) \\ \vdots \end{bmatrix} ; \quad \text{(b)} \quad {^\alpha}\mathbf{s} = \begin{bmatrix} {^\alpha}\mathbf{r}(t) \\ {^\alpha}\mathbf{v}(t) \\ {^\alpha}\sigma(t) \\ \vdots \end{bmatrix} \quad (3.7)$$

$$\text{(solids)} \qquad \qquad \text{(fluids)}$$

where in the form (a) $^i\sigma(t)$, $^s\sigma(t)$ refer to microstresses within an element α and on its boundary, respectively. The quantities $^i\mathbf{u}, {^s}\mathbf{u}$ are the corresponding deformations (see Chapter 4). The column vector pertaining to fluids indicated in (b) contains the configuration vector $^\alpha\mathbf{r}(t)$ or the velocity vector $^\alpha\mathbf{v}(t)$ of an element (molecule) and the microstress $^\alpha\sigma(t)$ that includes in general the contributions of interaction effects as well as other dynamic variables. Hence the following subspaces for the probabilistic analysis may be chosen:

$$^i\sigma(t) \in {^i\Sigma}, \quad {^s}\sigma(t) \in {^s\Sigma}; \quad {^i\Sigma} \oplus {^s\Sigma} = \Sigma \subset \mathscr{X}, \quad (3.8)$$

$$^i\mathbf{u}(t) \in {^iU}, \quad {^s}\mathbf{u}(t) \in {^sU}; \quad {^iU} \oplus {^sU} = \mathscr{U} \subset \mathscr{X}, \quad (3.9)$$

$$^\alpha\mathbf{r}(t) \in \mathscr{C}, \quad \mathscr{C} \subset \mathscr{X}; \quad {^\alpha}\mathbf{v}(t) \in \mathscr{V}, \mathscr{V} \subset \mathscr{X}, {^\alpha}\sigma(t) \in \Sigma, \Sigma \subset \mathscr{X}. \quad (3.10)$$

Evidently, for the representation of the physical system by an abstract dynamical one, the probability space $\mathscr{X}$ and any of its subspaces must be given a suitable structure, and an appropriate measure. Since by Postulate 2 of probabilistic mechanics all field variables are considered as random variables, the latter can be characterized by their probability distribution functions (see Chapter 1). It may be recalled that such a function in terms of the event sets will have the following properties:

$$\left.\begin{array}{l} \text{(i) } 0 \leqslant P\{E_n\} \leqslant 1 \text{ for } \forall E_n \in \mathscr{F}, \\ \text{(ii) } P\{E_1 \cup E_2\} = P\{E_1\} + P\{E_2\} - P\{E_1 \cap E_2\} \\ \qquad \text{if } E_1, E_2 \in \mathscr{F} \text{ and } E_1 \cap E_2 \neq \phi, \\ \text{(iii) } P\left\{\bigcup_r E_r\right\} = \sum_r P\{E_r\}, \text{ if } E_r \cap E_s = \phi, r \neq s, \\ \text{(iv) } P\{\mathscr{X}\} = 1. \end{array}\right\} \quad (3.11)$$

This shows that the distribution function $P\{E_n\}$ satisfies all the properties of a measure, i.e. that it is an extended real-valued, non-negative and countably additive set function. Hence introducing P as a probability measure on $[\mathscr{X}, \mathscr{F}]$ one can construct a probability space denoted by $[\mathscr{X}, \mathscr{F}, P]$. In view of the above definition of a measure and the corresponding probability space, the open Borel sets E defined before are taken as events of a random experiment. Thus the state of a microelement α within the specified range Def. 11 (iii) becomes an event $E(\mathbf{s})$ with a probability measure:

$$\mathscr{P}(\mathbf{s}_n) = \mathscr{P}\{E(\mathbf{s}_n)\} = \mathscr{P}\{\mathbf{s}_n < {}^\alpha \mathbf{s} < \mathbf{s}_n + \Delta \mathbf{s}_n\}; \quad \mathscr{P}\{\mathscr{X}\} = 1 \quad (3.12)$$

Considering the subspaces of $\mathscr{X}$ each of them will be measurable, if and only if E is the union of disjoint open sets $E_n, k = 1, \ldots, k$ (see Chapter 1). Thus one may consider, for instance, that part of E which relates to the deformations ${}^\alpha \mathbf{u}$ of an element only in a solid with microstructure. In this case the event will be specified by:

$$E(\mathbf{u}_n) = \{\mathbf{u}_n < {}^\alpha \mathbf{u} < \mathbf{u}_n + \Delta \mathbf{u}_n\}; \quad \bigcup_{n=1}^{\infty} E_n = \mathscr{X} \quad (3.13)$$

and the corresponding probability measure becomes:

$$\mathscr{P}(\mathbf{u}_n) = \mathscr{P}\{E(\mathbf{u}_n)\} = \mathscr{P}\{\mathbf{u}_n < {}^\alpha \mathbf{u}_n < \mathbf{u}_n + \Delta \mathbf{u}_n\}; \quad P\{\mathscr{U}\} = 1. \quad (3.14)$$

In probabilistic mechanics this measure is called the probability density of microdeformations. In most cases, however, the distribution functions only are experimentally accessible. It becomes then necessary to consider cumulative measures, which can be defined in terms of the distribution function of the above microdeformations as:

$$P^u = \mathscr{P}\{E(\mathbf{u}_n)\} = \mathscr{P}^u\{{}^\alpha \mathbf{u} < \mathbf{u}\}; \quad \mathscr{P}^u\{\mathscr{U}\} = 1. \quad (3.15)$$

Analogously, a cumulative measure can be given for any other field variable and generally using the state vector $\mathbf{s}$ one has:

$$\mathscr{P} = \mathscr{P}\{E(\mathbf{s})\} = \mathscr{P}\{{}^\alpha \mathbf{s} < \mathbf{s}\}; \quad \mathscr{P}\{\mathscr{X}\} = 1. \quad (3.16)$$

Since the above measure is closely related to the distribution function of probability theory, one can represent the probability space of micro-

3.4 The structure of the probabilistic function space $\mathscr{X}$

deformations in a solid or the velocity space in a discrete fluid by the abstract dynamical systems [$\mathscr{U}, \mathscr{F}^u, \mathscr{P}^u$] and [$\mathscr{V}, \mathscr{F}^v, \mathscr{P}^v$], respectively. It is to be noted, however, that the measures $\mathscr{P}^u, \mathscr{P}^v$, etc., must satisfy the regularity conditions. To show that $\mathscr{P}^u$, for instance, is regular, consider the σ-algebra $\mathscr{F}^u$ of the space $\mathscr{U}$. Since $\mathscr{F}^u$ contains the open sets $E(\mathbf{u}_n)$, it can be shown that the intersection of a countably finite number of open sets $E(\mathbf{u}_n)$ forms a closed set $\overline{E}(\mathbf{u}_n) \in \mathscr{F}^u$. Hence in accordance with the definition of regularity given in Chapter 1, section 1.2.3, $\mathscr{P}^u$ is also a measure on $\overline{E}(\mathbf{u}_n) \in \mathscr{F}^u$ so that $\mathscr{P}^u$ satisfies the condition of regularity. Hence the random deformation vector $^\alpha \mathbf{u} \in \mathscr{U}$ can be defined as a $\mathscr{P}^u$-regular measurable function in $\mathscr{U}$ (see also the definitions given by Doob [33] and Billingsley [113]).

The identification of [$\mathscr{U}, \mathscr{F}^u, \mathscr{P}^u$] as the space of all $\mathscr{P}^u$-regular measurable functions of $\mathbf{u}$ and similarly of [$\mathscr{V}, \mathscr{F}^v, \mathscr{P}^v$] as the space of all regular measurable functions of $\mathbf{v}$ with a probabilistic function space forms the basis for the mathematical structure of the probabilistic theory of deformation and flow of discrete media. In the case of solids this leads directly by recognizing the duality with the corresponding stress space (see section 3.6) to response relations in an operational form. Thus, following relation (3.9) concerning the deformations of a structured solid, the space $\mathscr{X}$ may be thought of being composed of the deformation space $\mathscr{U}$ and the stress space Σ. For a given solid a mapping between $\mathscr{U}$ and Σ will always exist, whereby the topological structure of one of these spaces, say Σ, may be given in terms of a non-degenerate bilinear form with respect to this mapping.

The above function space approach requires the use of either a set of norms or semi-norms in the associated probability spaces. For example, if $\mathscr{U}$ is the space of all $\mathscr{P}^u$-regular measurable and bounded functions of which are discrete, the expected value of $\mathbf{u}$ will be given by:

$$\langle (\mathbf{u}) \rangle = \Sigma \mathbf{u} \, \mathscr{P}\{E(\mathbf{u})\} = \Sigma \mathbf{u} \Delta \mathscr{P}^u. \qquad (3.17)$$

Thus in an experiment, where $\langle (\mathbf{u}) \rangle$ is equal to zero, although individual micro-deformations $^\alpha \mathbf{u}$ still exist, it is seen that $|\langle (\mathbf{u}) \rangle|$ is a semi-norm. In this case the deformation space will be a Fréchet space with this semi-norm. However, if in the same experiment one is rather interested in finding the average value of the magnitude of the microdeformations $\mathbf{u}$ a similar definition to relation (3.17) would satisfy the properties of a norm, i.e.

$$\|\mathbf{u}\| = \langle |\mathbf{u}| \rangle = \int_{\mathscr{U}} |\mathbf{u}| d\mathscr{P}^u$$

in which $\mathscr{U}$ is a Banach space. Using, for instance, the standard deviation of $\mathbf{u}$ one obtains in accordance with (3.17):

$$D^u = \{\Sigma |\mathbf{u}_s - \langle \mathbf{u}_s \rangle|^2 \Delta \mathscr{P}^u\}^{\frac{1}{2}} \qquad (3.18)$$

70 General Formulation of Probabilistic Mechanics

where D'' again satisfies the properties of a semi-norm. If $\mathcal{U}$ is a locally convex space with this semi-norm and if it is completely metrizable with respect to $\langle|\mathbf{u}|\rangle$ it is a Fréchet space. Although the analysis in a Fréchet space is analogous to that in a Banach space, the choice of the proper space will depend on the type of problem and the medium under consideration. These subspaces will be further considered in the probabilistic mechanics of solids and fluids.

3.5 INTERACTION EFFECTS—POTENTIALS

In section 3.3 of this chapter the fundamental concepts of probabilistic mechanics and axiomatic definitions of some of the relevant field quantities were given. It has been pointed out that the behaviour of a discrete medium can be characterized by a set of state vectors belonging to a probabilistic function space and an abstract dynamical system $[\mathcal{X}, \mathcal{F}, \mathcal{P}]$ where $\mathcal{F}$ is the σ-algebra $\mathcal{B}$ of Borel sets on $\mathcal{X}$. In this sense the principle of continuity is introduced by means of neighbouring points in the sets of $\mathcal{X}$. This concept also permits the inclusion of interaction effects between contiguous elements of the microstructure in the analysis.

Among the various interaction effects that may occur during deformation and flow of discrete media and which may be of a physical and chemical nature, the most significant in the present context are those due to interaction forces and couples. Such forces are in general derivable from potentials that are usually assumed to occur at discrete points or at clusters of such points on interfacial surfaces separating the elements of the microstructure. In a crystalline solid, for example, the term interface applies actually to an interaction zone of finite dimensions between neighbouring crystals. The existing bond-potential can be used in the analysis of bond-strength, cohesive and adhesive behaviour between different constituents of the solid and bonding characteristics of fibrous structures, etc. In liquids interaction effects are rather determined by short- or long-range forces, when models with central force potentials are introduced in the analysis of molecular flow. Here the term interaction region is more appropriate and will be discussed in more detail in Chapter 5.

(i) *Interatomic and intermolecular forces*

In the definition of a microelement (section 3.2), the meaning of a molecule and an intermolecular force has not been discussed in detail. However, for the molecular description of the response of discrete media, it becomes necessary to clarify these terms. A molecule can be regarded as a group of atoms (or a single atom), whose binding energy is large enough to permit it to interact with its surroundings without losing its structural identity. Thus, for instance, a hydrogen molecule can be classified as a molecule, whilst an argon molecule is

3.5 Interaction effects—potentials

better considered as a bound pair of argon atoms. If a molecule is lightly bound, its thermally populated vibrational and rotational states have similar structure and properties. In the case of argon, however, the structure and properties can vary considerably from state to state. Moreover, in a non-rigid molecule such as ammonia, for example, small changes in energy may be associated with a large change in an internal coordinate. In such cases, it is more significant to analyze the changes in conformation of the molecule with changes in its environment. It is well known that the equilibrium and non-equilibrium properties of matter are characterized by the presence of interatomic and intermolecular forces. Simple considerations of liquids clearly indicate, that molecules attract one another when they are sufficiently apart. The fact that liquids and solids have finite densities of a magnitude observed under normal conditions also shows that molecules repel at short distances. There exists therefore a balance point of zero force corresponding to a minimum energy. Hence the mutual potential energy of two atoms in a molecule or the interaction potential between two monatomic molecules held together by central forces must be generally of the form indicated by Fig. 1(a) below.

(ii) *Interaction potentials*

Figure 1(a) clearly indicates that the repulsive force dominates at small separations of atoms or molecules, whilst at larger separations an attractive

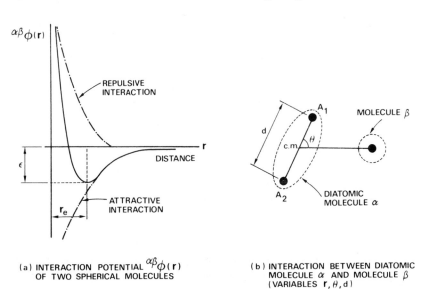

(a) INTERACTION POTENTIAL $^{\alpha\beta}\phi(r)$ OF TWO SPHERICAL MOLECULES

(b) INTERACTION BETWEEN DIATOMIC MOLECULE α AND MOLECULE β (VARIABLES r, θ, d)

FIG. 1. Schematic form of the pair-potential $^{\alpha\beta}\phi(\mathbf{r})$ in central force models.

force becomes predominant. The balance of these forces occurs at an equilibrium separation r_e that would conform to the actual distance between the atoms of a diatomic molecule. In a solid r_e will not necessarily be equal to the nearest-neighbour separation $^{\alpha\beta}r$ due to effects caused by the more distant neighbours, but it will be near to $^{\alpha\beta}r$, if the forces are assumed to be central. If additional non-central forces are present in the system, $^{\alpha\beta}r$ may be appreciably different from r_e or the latter may not exist at all. The existence of a potential energy function is based from statistical mechanics on the well-known Born–Oppenheimer approximation [64], whereby the nuclei in a particular configuration are fixed and the potential $\phi(r)$ is considered as the difference in energy of the system in that configuration from its value when the separation $r \to \infty$. It is to be noted that the number of variables on which the interaction energy depends increases sharply with an increase in molecular size. Thus, for instance, for an atom β and a diatomic molecule α as indicated in Fig. 1(b), there are three variables, i.e. the separation r of β from the centre of mass (c.m.) of α the angle θ between the molecular axis A_1–A_2 and the intermolecular separation r and the distance d between the nuclei A_1, A_2 in the molecule. For two diatomic molecules there are six variables, i.e. the intermolecular separation r, the angles θ_1, θ_2 between the molecular axes and the line of centres, an angle φ between the plane containing **r** and the molecules and two intramolecular distances d_1, d_2. In general for two interacting non-linear polyatomic molecules, the interaction energy will depend on $3(n_\alpha + n_\beta) - 6$ independent variables, where n_α, n_β denote the number of atoms in molecules α, β respectively. Six of these variables are required for the description of the relative position and orientation of the molecules α, β and the remaining $(3n_\alpha - 6) + (3n_\beta - 6)$ are internal vibrational coordinates of the two molecules. In certain cases these coordinates are of minor or no interest so that a considerable reduction of the variables can be achieved. By considering a fluid or gas on the basis of a hard-sphere model, the force in the collision of two monatomic molecules is given from physics to be equal to $-\partial \phi(r)/\partial r$ (Yvon [65]). In the case of a diatomic molecule interacting with an atom (Fig. 1(b)) there is also a torque present, which is equal to $-\partial \phi(r)/\partial \theta$. The above concept of interaction forces does not apply, however, for the determination of forces between ions in an aqueous solution. In such cases, it may be more convenient to use a potential of an average force, which is equivalent to the Helmholtz free energy representing the mean interaction energy of two ions at a fixed separation r averaged over all configurations of all other molecules and ions, that are present in the solution.

In considering intermolecular potentials one can distinguish two classes of interactions, i.e. short-range and long-range ones. The former decrease exponentially with increasing intermolecular distances r and are essentially due to an overlap of the electronic wave functions describing the isolated molecule. Long-range interactions vary in the form of r^{-m}, m being a positive

3.5 Interaction effects—potentials

integer, at large distances r. At long-range the electrons become rather indistinguishable in their exchange between molecules and hence in the determination of the interaction one can consider the electrons as if they were associated with one or the other of the molecules. Short-range forces may be attractive or repulsive, but at small separations, they are always repulsive. Long-range forces can also be attractive and repulsive. For spherical atoms the long-range forces are attractive. In this classification of short- and long-range interactions, the short-range group consists of overlap (Coulomb and exchange) forces that are non-additive. In the long-range interaction group, there are five distinct types of interaction energies and forces, e.g. electrostatic, induction, dispersion, resonance and magnetic. The electrostatic interactions are additive, the dispersion nearly additive and the magnetic only weakly so, the induction and resonance effects are, however, non-additive. For a more comprehensive study see, for instance, Hirschfelder [66], Margenau and Kestner [64], Buckingham [67, 68], Schuster [69] and others.

The intermolecular potential of simple liquids may be assumed to be the sum of effective pair-potentials in a first approximation. The latter for a real simple liquid is characterized by a short-range strong repulsion and a long-range attraction. More recently investigations concerning the theory of pair-potentials have become important as well as the conditions on these potentials for the existence of a solid–liquid transition. As classified above, the simple fluids are characterized by spherically symmetric non-saturating interactions. It will be further assumed in the analysis of simple liquids that the forces are central, i.e. acting through the centre of gravity of the molecule or particle and that they are pair-wise decomposable. In this sense the total N-body configurational energy can be represented by the sum of pair-interactions where the factor $\frac{1}{2}$ prevents counting the interactions $\alpha\beta$ and $\beta\alpha$ as distinct. Hence:

$$\phi(1,\ldots,N) = \frac{1}{2}\sum_{\alpha}^{N}\sum_{\beta}^{N}\phi(\alpha,\beta) = \sum_{\alpha>\beta}^{N}\phi(\alpha,\beta). \tag{3.19}$$

This relation holds strictly for two-body effects only and thus excludes three-body effects, since for the latter a correction term in the form of triplet potential $\phi(\alpha\beta\gamma)$ would have to be included in the summation $\phi(\alpha\beta)+\phi(\beta\gamma)+\phi(\gamma\alpha)$. This, however, is usually not done due to the large increase in the calculations of the total interaction potential. In dealing with simple liquids, it is usually assumed that the well-known Lennard-Jones potential and a number of idealized interactions (hard-spheres, square well, etc.) can be used in an approximation theory.

By considering crystalline solids the pair-potential of atoms α, β at separation r designated by $^{\alpha\beta}\phi(r)$ can also be taken as $\frac{1}{2}\phi(\alpha,\beta)$ such that the potential energy per atom can be expressed by a lattice sum $\Sigma\frac{1}{2}\phi(\alpha,\beta)$, which is taken over all pairs that an atom can form with the remaining ones in the lattice. For short-range interactions only the nearest neighbour of α need be included in this sum. If $\phi(r)$ is the only potential present in the structure of the solid, its form must be such that the lattice is mechanically stable and an energy increase can only be ascribed to an arbitrary small perturbation of the atoms from their equilibrium positions. Considerations of this type will be given in the following chapter, when dealing with grain boundary effects in crystalline solids. It is evident, that the conditions which $\phi(r)$ has to satisfy for a mechanically stable structure to exist are:

(i) it must have a minimum at a finite r,
(ii) for large separations, its magnitude must tend to zero more rapidly than r^{-3}.

These conditions and the finiteness of the lattice energy ensure the stability of the structure with respect to an infinitesimal homogeneous tension or compression of the lattice. However, shear deformations are of equal importance for lattice stability, but this will not be further pursued here and reference is made to the work of Born [70], Mott [71] and others. A frequently used form of $^{\alpha\beta}\phi(r)$ in crystalline solids is:

$$^{\alpha\beta}\phi(r) = A\left(\frac{r_e}{r}\right)^n - B\left(\frac{r_e}{r}\right)^m, \quad n > m \qquad (3.20)$$

in which the first term represents the repulsive part of the potential and the second the attractive part, r_e the equilibrium separation and A, B are material characteristics. So far as central force models are concerned, it is known that for crystals of the solidified rare gases, the attractive part of the potential originates from van der Waals forces. It can be shown, that m in (3.20) is equal to six. The most appropriate repulsive potential corresponds to values of $n = 10 \div 12$. If 12 is chosen the interactions are sometimes called Lennard-Jones forces and the interaction potential the L.J.-potential [72].

Although no specific form of the pair-potential can be given for the great variety of polycrystalline solids and in particular for metals, it is frequently assumed that the L.J.-potential is a good approximation. An alternative way to the assumption of a force law consists of evaluating the parameters in (3.20) by the comparison with experimentally obtained values for a given solid. It is sometimes preferable to use instead of (3.20) a potential in which the repulsive and attractive parts are expressed in form of exponentials. In this case the potential is given by:

$$^{\alpha\beta}\phi(r) = D\{\exp[-2v(r-r_e)] - 2\exp[(-v(r-r_e)]\} \qquad (3.21)$$

where D and v are material characteristics established by appropriate experiments. This form is known as the Morse potential and has been found useful in the description of the behaviour of both b.c.c. and f.c.c. metals for instance. It has been applied in a modified form in the study of interaction effects in polycrystalline solids and fibrous structures (ref. [36]). By introducing instead of the hard-sphere model a soft-sphere one, the corresponding L.J.-potential can be extended to contain another term in relation (3.21). Such a modification has been suggested by Matsuda and Hiwatari (see ref. [73]) for the study of solid–liquid phase-transitions.

3.6 DUALITY OF SPACES

The formulation of a general stochastic deformation theory of structured solids employs an operational formalism. This formalism is well known in mathematical physics and is often used in the theory of random equations. This type of equation has become only recently the subject of intensive research. Contributions to this field are due to Bharucha-Reid [74], Itô [75], Arnold [76], Hanš [77], Špaček [78], McKean [79] and others. In the application of such an operational formalism to the stochastic mechanics of solids, it becomes necessary to introduce a material functional or operator. This has already been indicated by Postulate 4 of section 3.3. The notion of a material operator or linear functional is of great importance in the probabilistic theory of solids. The operator reflects on the one hand the material characteristics of a specific solid, and on the other, it allows for the use of the duality of the stress and deformation space. In general the duality of linear vector spaces plays an important role in convex analysis. The mathematical structure of a pair of linear spaces placed in duality by a bilinear form is equivalent to the classical method of mechanics using the virtual work principle.

(1) Duality in Banach spaces

Since most of the field quantities in the formulation of the mechanical behaviour of structured solids are elements of a Banach space, the concept of duality in such spaces will be briefly considered. The earlier given definitions of Banach space-valued random variables and functions will be used for the subsequent definition of the duality of such spaces. Considering the probabilistic function space $[\mathscr{X}, \mathscr{F}, \mathscr{P}]$, assumed to be a complete probability measure space, and a measurable Banach space $[X, \mathscr{A}]$, where $\mathscr{A}$ is the σ-algebra of Borel subsets of X, a Banach space-valued random variable is then defined by the mapping $x: \mathscr{X} \to X$, if $x^{-1}(A) \in \mathscr{F}$ for all $A \in \mathscr{A}$. This definition makes the Banach space-valued random variable x a Borel measurable function in the same way as in the classical probability theory. Indeed, if X is the real line $\mathbb{R}^1$, for example, or if $X \equiv \mathbb{R}^n$ (Euclidean n-space) the above definition of X becomes identical to that of an ordinary random variable. There are various other definitions of Banach space-valued random variables, which were given by Fréchet in [80], Mourier [81], Hanš [82] and others. If the space X is separable, however, these definitions become equivalent. The property of separability of a Banach space is also significant from another point of view, e.g. if the field quantities in probabilistic mechanics can only be defined in terms of semi-norms. This leads then to the notion of a locally convex space, which is defined by a system of semi-norms satisfying the property of separation. If there exists only one semi-norm, the corresponding linear space becomes a normed linear space. A Banach space X is separable, if it has a countable subset that is everywhere dense (Kuratowski [4]). Separable Banach spaces are an important class of Banach spaces in their application in probability theory. The $\mathbb{R}^n$ or Euclidean n-space is most significant in the mechanics of solids. It is the space of real numbers $x = (x_1, \ldots, x_n)$ and a real Banach space with respect to coordinatewise addition and scalar multiplication. It has a norm given by:

$$\|x\| = \left\{ \sum_{i=1}^{n} |x_i|^2 \right\}^{1/2}. \tag{3.22}$$

In order to clarify the concept of duality in Banach spaces, the notion of a linear functional (Chapter 1) on a Banach space X is considered next. It is a function on X to the scalars. Thus denoting by $f(x)$ the value of f for an element $x \in X$ the functional f will be linear if:

(i) $f(x_1 + x_2) = f(x_1) + f(x_2); \quad x_1, x_2 \in X;$

(ii) $f(\alpha x) = \alpha f(x); \quad x \in X, \alpha = \text{scalar}.$ \quad (3.23)

The functional f is said to be bounded, if there exists a real constant $M \geq 0$ such that:

$$|f(x)| \leq M \|x\| \quad \text{for all} \quad x \in X.$$

A linear functional is bounded, if and only if it is continuous. Denoting the set of all bounded linear functionals on a Banach space X by X^*, the latter is also a Banach space or the dual of X. Since X^* denotes all bounded linear functionals on X an element of X^* can be designated by x^*. Thus, $x^*(x)$ gives the value of x^* for an element $x \in X$. Similarly, forming the set of all bounded linear functionals on X^* gives the second dual or adjoint space of X written as X^{**}. If $X = X^{**}$ the Banach space is said to be reflexive. It is convenient to designate $\langle x, x^* \rangle$ instead of $x^*(x)$, where this notation indicates the duality, which exists between X^{**} and X as well as the action of X on X^*. The following characteristics of duality can be given. Suppose B is a closed unit sphere of a normed space, then the norm of the element x^* is given by:

$$\|x^*\| = \sup\{|\langle x, x^* \rangle| : x \in B\} \quad \text{for every} \quad x^* \in X^* \tag{3.24}$$

and hence

(i) this norm makes X^* a Banach space,
(ii) if B^* is a closed unit sphere of X^*, then for every $x \in X$,
$$\|x\| = \sup\{|\langle x, x^*\rangle| : x^* \in B^*\}.$$
Hence, $x^* \to \langle x, x^*\rangle$ is a bounded linear functional on X^* with norm $\|x\|$.
(iii) B^* is called weak*-compact. (3.25)

Evidently, several topologies can be introduced in Banach spaces, each of which will be associated with a certain type of convergence of the random variable. For instance, if the topology induced in X occurs by the metric (distance function) $d(x, y) = \|x - y\|$, it is called a metric norm or a strong topology of X. A weak topology in X is induced by a sequence $\{x_n\}$ of elements in a Banach space converging in a weak topology to an element x, if

(i) the norms $\|x_n\|$ are uniformly bounded, i.e. $\|x_n\| \leq M$,
(ii) $\lim_{x_n \to x} x^*(x_n) = x^*(x)$ or $\langle x^*, x\rangle$ for every $x^* \in X^*$. (3.26)

The weak topology when introduced in the dual space X^* is called a weak*-topology of X^*.

(2) *Operators on Banach spaces—material operators*

It has been stated earlier that for convenience of the analysis one can consider subspaces of the probabilistic function space $\mathscr{X}$. Since these spaces are regarded as subspaces of a general Banach space the mapping between them should be clarified. In particular, since in the definition of the field variables three measuring scales have been introduced, i.e. the micro-, meso-, and macroscopic scale, the corresponding operators mapping the stress into the deformation subspaces have to be defined. These operators are designated as material operators. In general a macroscopic material operator reflects the material characteristics of a specific structured solid and can be formally expressed by:

$$\mathscr{M} = \mathscr{M}\{\underline{E}, \mathbf{a}, \rho, \phi, \mathbf{O}, \Theta, t, \dots\} \quad (3.27)$$

where $\underline{E}$ is the modulus of elasticity of the material, $\mathbf{a}$ the lattice parameter, ρ the dislocation density, ϕ an interaction potential, $\mathbf{O}$ an orientation matrix, Θ the temperature, t the time, etc. It is, of course, extremely difficult to assess all parameters affecting the material response characteristics. However, only those parameters need be considered that are significant in the particular phenomenon under consideration.

In order to discuss the type of operators involved in the analysis and

78 General Formulation of Probabilistic Mechanics

beginning with the smallest scale, i.e. that of a microelement, one can define a local material operator or microelement operator as follows:

Def. 1:

$$\left.\begin{array}{c} {}^\alpha m:\ {}^m\Sigma_d \to {}^m \mathcal{U}_d;\ {}^{\mathcal{D}_m}\Sigma_d \oplus {}^{\partial \mathcal{D}_m}\Sigma_d = {}^m\Sigma_d \subset \mathcal{X} \\ \text{for all } \alpha = 1, 2, \ldots, N \end{array}\right\} \quad (3.28)$$

where the superscripts $\mathcal{D}_m$, $\partial \mathcal{D}_m$ indicate the domain of the microelement and its boundary, respectively. The subscript d refers to dense sets in the stress space $\Sigma \subset \mathcal{X}$ and deformation space $\mathcal{U}$ respectively.

On the assumption that these microelement operators are linear and bounded, the inverse of the operators can also be used so that:

$$ {}^\alpha m^{-1}: {}^m\mathcal{U}_d \to {}^m\Sigma_d;\ {}^{\mathcal{D}_m}\mathcal{U}_d \oplus {}^{\partial \mathcal{D}_m}\mathcal{U}_d = {}^m\mathcal{U}_d \subset \mathcal{X} \quad (3.29)$$

and hence the microdeformations can be related to the microstresses by:

$$ {}^\alpha \mathbf{u} = {}^\alpha m^\alpha \sigma;\ {}^\alpha \sigma \in {}^m\Sigma_d,\ {}^\alpha \mathbf{u} \in {}^m\mathcal{U}_d \quad (3.30)$$

in which $^\alpha\sigma$ is in general a generalized microstress (see also references [36]), acting on a microelement and $^\alpha\mathbf{u}$ the corresponding deformation.

Considering the next scale, i.e. a mesodomain of the material body containing an ensemble of microelements, the corresponding operator is then defined by the expected value of the micro-operators as follows:

Def. 2:

$$M = E\{{}^\alpha m\} = \sum_{\alpha=1}^{N} {}^\alpha m\, \Delta\, \mathcal{P}\{{}^\alpha m\} \quad (3.31)$$

for a specific meso-domain $\mathcal{D}_M$. Again such an operator and its inverse will have to be considered for the interior of a mesodomain $\mathcal{D}_M$ and on its boundary $\partial \mathcal{D}_M$ so that:

$$\left.\begin{array}{c} M: {}^M\Sigma \to {}^M\mathcal{U};\ {}^{\mathcal{D}_M}\Sigma \oplus {}^{\partial \mathcal{D}_M}\Sigma = {}^M\Sigma \subset \mathcal{X};\ M = 1, 2, \ldots, P \\ \text{and} \\ M^{-1}: {}^M\mathcal{U} \to {}^M\Sigma;\ {}^{\mathcal{D}_M}\mathcal{U} \oplus {}^{\partial \mathcal{D}_M}\mathcal{U} = {}^M\mathcal{U} \subset \mathcal{X}. \end{array}\right\} \quad (3.32)$$

It is evident that for the required macroscopic formulation of the material operator connecting the macroscopic stress space $^\mathcal{M}\Sigma$ to the macroscopic deformation space $^\mathcal{M}\mathcal{U}$ an analogous definition will apply. Thus the macroscopic material operator can be regarded as the mapping:

Def. 3:

$$\mathcal{U}: {}^\mathcal{M}\Sigma \to {}^\mathcal{M}\mathcal{U};\ {}^\mathcal{M}\Sigma, {}^\mathcal{M}\mathcal{U} \subset \mathcal{X} \quad (3.33)$$

3.7 Operational forms of macroscopic constitutive relations   79

in which $^u\Sigma$, $^u\mathcal{U}$ are recognized to be the collections of the meso-scopic subspaces $^M\Sigma_p$, $^M\mathcal{U}$ respectively. Thus

$$^{\mathcal{M}}\Sigma = \bigoplus_{p=1}^{P} {^M\Sigma_p} \quad \text{and} \quad {^\mathcal{M}\mathcal{U}} = \bigoplus_{p=1}^{P} {^M\mathcal{U}_p} \tag{3.34}$$

It is seen that the macroscopic material operator $\mathcal{M}$ is an element of the space of linear continuous operator, i.e.:

$$\mathcal{M} \in \mathscr{C}(\cdot^{\mathcal{M}}\Sigma, {^\mathcal{M}\mathcal{U}}). \tag{3.35}$$

The above subspaces of $\mathscr{X}$ are assumed to be Banach spaces of the same type defined as both real or complex. In general, $\mathscr{X}$ will be a Banach space, if the proper norm can be introduced for its topology. However, as mentioned before, it may not always be possible to introduce such a norm and hence seminorms may have to be employed. In that case the analysis will be similar and conform to one pertaining to Fréchet spaces. The quantities belonging to the microelement state vector $^\alpha$s of the structure such as microstress tensor $^\alpha\boldsymbol{\sigma}$ and microdeformation vector $^\alpha\mathbf{u}$ render the local material operator in an explicit form a third-rank tensor. Some forms of such operators have been given previously on the basis of microkinematic studies in references [36, 83, 84, 85].

3.7 OPERATIONAL FORMS OF THE MACROSCOPIC CONSTITUTIVE RELATIONS

It has been stated earlier, that in lieu of the conventional constitutive relations of continuum mechanics, operational forms of such relations are employed in stochastic mechanics. These operators are based on the above definitions and are subject to the theorems of operator theory. Their explicit form will depend mainly on the choice of a proper microelement of the structure and also on the form of interaction between such elements. In general, deformations occurring in a structured solid can be regarded as stochastic processes, in which the operators map at any instant of time elements of the stress space into elements of the deformation space and hence perform the basic role of constitutive relations.

The macroscopic material operator in general is also time-dependent and hence will be designated by $\mathcal{M}(t)$. From the statement given in (3.35) this operator evidently connects the macroscopic stress and deformation fields induced in the macroscopic material body by external loads. It should be noted that there are certain restrictions on this operator such as measurability, convergence, etc. In applications it is convenient to consider the macroscopic stress–deformation relations in the form of the distribution of the mesoscopic operators. Hence the operator $M(t)$ can be employed and the evolution of its distribution with time during deformations may be obtained [2, 86]. If for the

80 General Formulation of Probabilistic Mechanics

representation of the deformation process a time-homogeneous Markov process is employed, the semi-group analysis shown earlier can be used. Considering the micro-deformations $\mathbf{u}(t)$ as statistically independent random variables and using the inverse operator $M^{-1}(t) \in \mathscr{C}(^M\mathscr{U}, {}^M\Sigma)$ and if an appropriate measure on this space will be introduced, the following relation between the corresponding distributions will exist:

$$\mathscr{P}\{\boldsymbol{\sigma}(t)\} = \mathscr{P}\{M^{-1}(t)\}\,\mathscr{P}\{\mathbf{u}(t)\}. \tag{3.36}$$

Assuming that the process of deformation is a time-homogeneous one and that the microstress distribution with time in the stress space ${}^M\Sigma \subset \mathscr{X}$ is Markovian, it has been shown in ref. [88] that the stress distribution is given by:

$$\mathscr{P}\{\boldsymbol{\sigma}(t)\} = \underline{P}^\sigma(t)\,\mathscr{P}\{\sigma(0)\} \tag{3.37}$$

in which the matrix $\underline{P}^\sigma(t)$ is identified with the transition probability of a Markov process in the stress space so that the Kolmogorov differential equation for the stress is as follows:

$$\frac{\mathrm{d}}{\mathrm{d}t}\underline{P}^\sigma(t) = \underline{Q}^\sigma(t)\,\underline{P}^\sigma(t);\quad \underline{P}^\sigma(0) = \underline{I} \quad \text{(identity matrix)}. \tag{3.38}$$

This relation represents the evolution of the stress distribution with time. An analogous equation will evidently exist for the evolution of the distribution of microdeformations in the deformation space ${}^M\mathscr{U} \subset \mathscr{X}$.

In the case where the probabilistic and topological structure of the stress or deformation space is unknown or the Markovian character cannot be assumed, it is still possible to construct non-degenerate bilinear forms with respect to the inverse operator $M^{-1}(t)$ (see also Moreau [87]) so that $M^{-1}(t)$ can be used.

Considering further the Markovian process, then by using the inverse of the operator, i.e. $M^{-1}(t)$ or its distribution and employing (3.33), which is valid for any time instant from $t = 0$ onwards, the following relation can be established:

$$\underline{P}^\sigma(t)\,\mathscr{P}\{\boldsymbol{\sigma}(0)\} = \mathscr{P}\{M^{-1}(t)\}\,\underline{P}^u(t)\,\mathscr{P}\{\mathbf{u}(0)\}. \tag{3.39}$$

In order that equation (3.35) is satisfied for all instants of time one obtains for $t = 0$ the following expression:

$$\mathscr{P}\{\boldsymbol{\sigma}(0)\} = \mathscr{P}\{M^{-1}(0)\}\,\mathscr{P}\{\mathbf{u}(0)\} \tag{3.40}$$

which together with (3.38) yields:

$$\underline{P}^\sigma(t)\,\mathscr{P}\{M^{-1}(0)\}\,\mathscr{P}\{\mathbf{u}(0)\} = \mathscr{P}\{M^{-1}(t)\}\,\underline{P}^u(t)\,\mathscr{P}\{\mathbf{u}(0)\}. \tag{3.41}$$

It is seen that, if the initial distribution of microdeformations $\underline{P}\{\mathbf{u}(0)\}$ is specified from experimental observations (see Chapter V of ref. [36]), the above relation leads to the following form connecting the stress and deformation distributions by means of the material operator $M^{-1}(t)$:

$$\underline{P}^\sigma(t)\,\mathscr{P}\{M^{-1}(0)\} = \mathscr{P}\{M^{-1}(t)\}\,\underline{P}^u(t). \tag{3.42}$$

3.7 Operational forms of macroscopic constitutive relations

It is seen that this leads to the following evolution-operator of $M^{-1}(t)$:

$$\mathcal{A}L(t)\mathcal{P}\{M^{-1}(0)\} \stackrel{df}{=} \mathcal{P}\{M^{-1}(t)\} = \underline{P}^\sigma(t)\mathcal{P}\{M^{-1}(0)\}(\underline{P}^u(t))^{-1} \quad (3.43)$$

It is to be noted that, if $P^\sigma(t)$ and $P^u(t)$ have the semi-group property, the evolution operator $\mathcal{A}L(t)$ will also have this property, i.e.:

$$\mathcal{A}L(t+h) = \mathcal{A}L(t)\,\mathcal{A}L(h)$$

Furthermore on the basis of relation (3.38), it can be readily shown that the following will hold:

$$\frac{d}{dt}\mathcal{A}L(t) = \underline{Q}^\sigma(t)\cdot\mathcal{A}L(t) - \mathcal{A}L(t)\underline{Q}^u(t) \quad (3.44)$$

As mentioned earlier, of the two possible ways to employ the material operator, the procedure of using the inverse of operator $M^{-1}(t)$ is preferable in applications since microdeformation distributions are experimentally accessible and the distribution of the material operator can be constructed from appropriate modelling of the microstructure. Examples of this procedure have been given in several publications ([88, 35, 89]). The above developed evolution of the material operator $M^{-1}(t)$ shows clearly the importance of the use of probabilistic functional analysis in the mechanics of discrete media (see also [36]).

4
Probabilistic Mechanics of Solids

4.1 INTRODUCTION

In the application of functional analysis on the basis of the probabilistic mechanics principles to the behaviour of solids, the microstructural characteristics of the latter have to be considered. As mentioned earlier, the structures of solids show two distinct features, i.e. a large number of internal surfaces, at which interactions between elements can occur and a variation in their geometrical and physical properties.

Interaction effects manifest themselves according to strong or weak interactions and depend largely on the type of microstructure. Thus, for instance, in a closed-packed arrangement (polycrystalline solids), the interaction effects relate to bonding and fracture of the solids. Models of bonding and fracture in the probabilistic formulation will be considered in section 3 of this chapter. Section 4.4 is concerned with the development of a general probabilistic deformation theory and the microstructural stability. The concept of material operators given in Chapter 3 is important for the establishment of constitutive relations and boundary value problems of solids. It will be discussed in section 4.6. The last section of this chapter deals with the probabilistic mechanics formulation of the wave propagation in structured solids (section 4.7).

Due to the great variety and complexity of microstructures in solids an attempt has been made to classify them in previous work [36]. In this study only two types of structures will be considered, i.e. polycrystalline solids and fibrous materials, since they are frequently encountered in practice.

4.2 ELEMENTS OF THE MICROSTRUCTURE IN SOLIDS

It is evident that the complexity of the geometrical arrangement of microstructures requires some simplification or idealization in order to consider them in the form of representative models in the analysis. Due to the limited scope of this text, the two characteristic groups mentioned above are briefly considered here. A schematic representation of such microstructures is given in Fig. 2(a, b).

4.2 Elements of the microstructure in solids 83

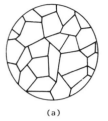
(a)
POLYCRYSTALLINE SOLID

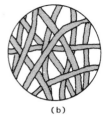
(b)
FIBROUS NETWORK

FIG. 2. Schematics of microstructures.

It can be seen from Fig. 2(a) that the great number of interfaces will play a significant role in the response behaviour of the polycrystalline solid and particularly in the breakdown of the structural arrangement. The arrangement shown in Fig. 2(b) for fibrous networks indicates that the interaction effects become important at the crossings of overlapping fibers. This may lead, through partial or complete bonding failure, to the complete failure of the structure. It is evident from these structural arrangements that the choice of an appropriate microelement becomes important. Thus, for the first structure obviously a single grain or crystal can be taken as a microelement. In the second example, as shown in previous work [90, 86, 91, 92], it is convenient to choose a microelement to consist of a free fiber segment and two halves of an effective bonding area between crossing fibers. Other characteristic structural arrangements are shown in Fig. 3. Thus, for instance, a composite material consisting of hard metal particles embedded in a ductile matrix is indicated in Fig. 3(a) whilst Fig. 3(b) represents a granular media.

It is evident, that the structures indicated below will have different interaction effects from the microstructures shown in Fig. 2(a, b). Hence, the corresponding material operators, which replace the conventional constitutive relations in the operational formalism of probabilistic mechanics, will have to reflect these differences. In this context, some material operators and their

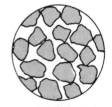

(a) COMPOSITE MATERIAL   (b) GRANULAR MATERIALS

FIG. 3. Other types of microstructures.

explicit forms for the two types of microstructures (Fig. 2(a, b)) have been given in previous work [36, 92].

The microelements of the structure are taken in accordance with the spatial scales in probabilistic mechanics as the primitive base of random functions in the analytical formulation. However, the microelements themselves have a structure at a smaller scale than that adopted in the subsequent analysis. Hence, in general, for the simplification of the formulation, the microelements in a first approximation can be regarded as continua. It is apparent that, if at this smaller scale existing defects are taken into account, the micromaterial operator will be of a more complex form. For instance, it has been shown in references [36, 94] that line defects, such as dislocations, can be built into such a material operator for the response of a single microelement in a polycrystalline solid. This leads, then, by considering also grain boundary effects between microelements to an overall response behaviour in such solids with the inclusion of microstructural effects. Similarly, one can include other defects of single microelements due to lattice imperfections, but not without increasing the complexity in the formulation. Similar consideration can be given to other material structures, which in certain instances require the recognition of internal effects within a microelement or cluster of such elements as, for instance, in the mechanics of soils.

4.3 INTERACTION EFFECTS

(i) *Effects of grain boundaries in polycrystalline solids*

Interaction effects between the elements of the microstructure in polycrystalline solids occur essentially on the grain boundaries, whilst in fibrous structures these effects depend on the bonding behaviour between individual fibers of the structure. Evidently, there will be analogous effects in other types of structures, which are not considered here. As mentioned before, the internal behaviour due to dislocations and other defects in single crystals as well as grain boundaries effects have been discussed in reference [95, 96]. For the consideration of interaction effects, the interfacial potential in the latter reference has been chosen as a Morse-type potential of the following form:

$$\phi\{|^{\alpha\beta}\mathbf{d}|\} = \phi_0\{1 - \exp[-b|^{\alpha\beta}\mathbf{d}|]\}^2 \qquad (4.1)$$

where the distance vector is given by:

$$^{\alpha\beta}\mathbf{d} = |^{\alpha\beta}\mathbf{d}|\mathbf{e}. \qquad (4.2)$$

In the adopted notation of the present analysis, the argument in the potential is taken as a relative displacement between coincident points in the lattices of the α, β crystals in accordance with Bollmann's [97] coincidence lattice theory of polycrystalline solids, and $\mathbf{e}$ is the unit vector in the direction of the relative

4.3 Interaction effects 85

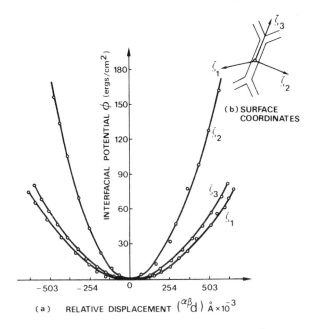

FIG. 4. Interfacial potential ϕ of copper (ergs/cm^2).

displacement. The interfacial potential ϕ for polycrystalline solids is considered as a surface potential occurring at the grain boundary, which is derivable from the Morse-function type of potential by a computer simulation technique. A plot of such an interfacial potential for copper, for example, [98, 83] is shown in Fig. 4. It is to be noted that to account for the mismatch angle between individual crystals and the boundary effect between them, it is convenient to use a surface coordinate system within the interaction zone for α and β (Fig. 4(b)). Thus, if $^\alpha\mathbf{n}$ denotes the normal to the surface of the αth crystal, then in terms of the surface coordinates one obtains the following relations:

$$^\alpha\zeta: {^\alpha\zeta_1} = {^\alpha\mathbf{n}} \times {^{\alpha\beta}\lambda}, \quad {^\alpha\zeta_2} = {^\alpha\mathbf{n}}, \quad {^\alpha\zeta_3} = {^{\alpha\beta}\lambda} \qquad (4.3)$$

in which the rotation about a common axis is expressed by the eigenvectors λ as indicated in Fig. 4(b) (see also reference [36]). The relative displacement vector in Fig. 4, $^{\alpha\beta}\mathbf{d}$ ($\mathring{A} \times 10^{-3}$), and the corresponding potential are indicated by three curves corresponding to the surface coordinate frame of a coincidence cell point at the surface of the lattice. The results of the computer simulation employed for the determination of grain boundary influences including thermal effects on grain boundaries have also been discussed in references [95, 96]. The two scalar quantities ϕ, ϕ_0 as well as the material characteristic b in relation (4.1) can be obtained from spectroscopic studies. The effect of the presence of the grain boundaries in the response behaviour of polycrystalline

86 Probabilistic Mechanics of Solids

solids particularly with respect to their elastic response has been considered in reference [36].

(ii) *Effect of bonding in fibrous networks*

The interaction effects between elements of a fibrous network such as polymers, cellulose structures, etc., can be illustrated by considering the bond behaviour of overlapping fibers discussed in earlier work [36, 99, 100, 103, 104]. Here again, the significant parameter is $^{\alpha\beta}\mathbf{d}$. This parameter, although it has the same interpretation as in the coincident cell model for crystalline solids, is taken here as a relative motion between matching points of two bonded unit cells in the fibers α, β, respectively. An illustration of this bond model is indicated by Fig. 5. It is seen from diagram (a), that schematically shows a typical scanning electron microscope observation, that two overlapping fibers will ideally bond over a certain area at their surface layers (Fig.

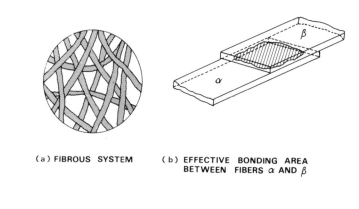

(a) FIBROUS SYSTEM   (b) EFFECTIVE BONDING AREA BETWEEN FIBERS α AND β

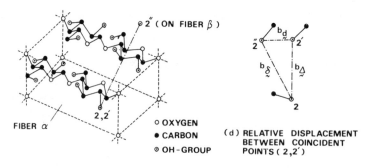

(c) UNIT CELL MODEL

FIG. 5. Bond-model (hydrogen bonding, reference [36]).

4.3 Interaction effects

5 (b)). In these surface layers the bonded area consists of a number of individual bonds belonging to it and result in an effective bonding area.

It is estimated that in a cellulose structure, for instance, there exist six hydroxyl groups per unit cell in such an area, which provide a perfect bonding between the fibers. The subdivision of the bonding area into such unit cells as indicated in Fig. 5(c) shows clearly that the arrangement refers to the atomic level of a hydrogen bond. The relative motion of such a bond in terms of the displacement of coincidence points (2, 2') is indicated in Fig. 5(d) by assuming a simple uniaxial force application. This model has been extended to allow for the occurrence of bending and shear modes in the bond deformation. To account for the interaction effect between the fibers, one can also use a Morse-type potential, which involves the relative displacement associated with an equilibrium position during the deformation of two bonded fibers, such that:

$$\phi(t) = \phi_0 \{\exp(-2v|^{\alpha\beta}\mathbf{d}(t)|) - 2\exp(-v|^{\alpha\beta}\mathbf{d}(t)|)\} \quad (4.4)$$

where ϕ_0 is an equilibrium potential, v a material characteristic of the fibrous structure and $^{\alpha\beta}\mathbf{d}(t)$ the relative displacement (Fig. 5(d)). It is perhaps of interest to note that the above form is only an approximation to the more rigorous bond potential, since the terms for the repulsion energy between the hydrogen and free oxygen atoms and that due to the exchange energy of attraction between the oxygen atoms have been neglected. However, the approximation from a probabilistic mechanics point of view will correspond to the adopted microscales (postulate (1)). It is to be noted that, in general, the interaction effects will be time-dependent. Hence, the models suggested above and as indicated by the form of potential in (4.4) have to be extended to include the mechanical relaxation of such materials. In the case of polycrystalline solids, a study to this effect and the corresponding formulation has been given in references [36, 101, 98]. It has been shown that the relaxation of a single crystal can be formulated in terms of internal stresses and deformations as well as surface stresses and surface deformations, respectively. By the use of an operational formalism the corresponding constitutive relations for creep and relaxation have been established. Similar considerations were given to the rheological behaviour of fibrous structures [90, 103, 104].

(iii) *Probabilistic models of fracture and bond failure*

In references [100, 105, 106] concerned with probabilistic models of polycrystalline solids, dislocation effects and the influence of grain boundaries have been discussed in some detail. In the present study it is of interest to consider probabilistic models of bond failure and fracture in such materials. Thus, so far as internal bonding in the lattice structure is concerned, missed bonds can be formulated by adopting a random-walk model, that accounts for missing bonds in a regular lattice (see, for instance, Montroll and

West [107]). However, on the assumption that a regular lattice is continued into a neighbouring crystal, i.e. by regarding the grain boundary as an interface only, between two regular lattices, such an approach can be used for the formulation of the bonding between crystals α, β.

It is usually assumed in the fracture mechanics of polycrystalline solids that a fracture zone consists of two parts. A cohesive zone, in which the neighbouring crystals still act as completely bonded and a free zone, in which bonding has ceased to exist. From this point of view, the free zone is one where the initial crack of a certain length a_0 will initiate a crack propagation towards, the cohesive zone. These two zones are characterized in continuum mechanics by smooth non-intersecting curves $z(s)$ of a certain arc length s, so that both these zones can be described in terms of time-dependent parameters as indicated in Fig. 6.

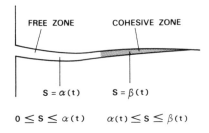

FIG. 6. Continuum model of crack propagation.

An analytical model corresponding to this picture of crack initiation and propagation in the continuum mechanics sense is due to Gurtin [108]. In contrast, from a probabilistic point of view, it is assumed that the initial crack formation is a decohesion phenomenon and the picture of crack initiation and propagation in a polycrystalline solid is rather seen as indicated in Fig. 7. It is evident that the crack, starting most commonly at a notch or any other surface defect at an open surface of the material body, will subsequently propagate randomly and that the process can be modelled by a Markov process. Hence the involved transition probability functions of the process and corresponding transition intensities will characterize generally a non-homogeneous birth–death Markov process. Such a formulation includes the participation of single microelements or grains and their respective number of bonds on their surfaces at a given time t. Analytically, this can be expressed in terms of the number of bonds per interfacial surface area $^\alpha a$ between α, β or internal in α.itself, where $^\alpha N(t)$: $\{^\alpha N(t), t \geqslant 0\}$. Thus considering a Markov chain, it will have the transition probability of the form:

$$p_{m,n}(s,t) = P\{^\alpha N(t) = n | ^\alpha N(s) = m\}. \tag{4.5}$$

The above transition probability can be expressed in terms of transition

4.3 Interaction effects

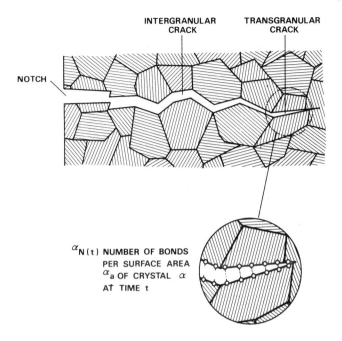

FIG. 7. Discrete model of crack propagation.

intensities on the assumption that the following non-negative functions exist, i.e.:

$$\lambda_0(t), \lambda(t), \ldots \quad \text{and} \quad \mu_1(t), \mu_2(t), \ldots \quad (4.6)$$

with the following limit conditions to hold at each time instant t, uniformly in n:

$$\left.\begin{array}{l}
\text{(i) } \lim_{\Delta t \to 0} \dfrac{p_{n,n+1}(t, t+\Delta t)}{\Delta t} = \lambda_n(t) \quad \text{for} \quad n \geq 0 \\[1em]
\text{(ii) } \lim_{\Delta t \to 0} \dfrac{p_{n,n-1}(t, t+\Delta t)}{\Delta t} = \mu_n(t) \quad \text{for} \quad n \geq 1 \\[1em]
\text{(iii) } \lim_{\Delta t \to 0} \dfrac{1 - p_{n,n}(t, t+\Delta t)}{\Delta t} = \lambda_n(t) + \mu_n(t), \quad n \geq 0
\end{array}\right\} \quad (4.7)$$

for all values of $t \geq 0$ and $\mu_0(t) = 0$. Thus, a pure birth process is then characterized by:

$$\left.\begin{array}{l}
\mu_n(t) = 0 \quad \text{for all } t \text{ and } n, \\
\text{and a pure death process by:} \\
\lambda_n(t) = 0 \text{ for all } t \text{ and } n.
\end{array}\right\} \quad (4.8)$$

In this manner one obtains the Chapman–Kolmogorov differential equation

(Chapter 2) for the evolution of the probabilities. In practical applications, it is more convenient, however, to use a generating function for the transition probabilities. Thus the crack formation and propagation based on the decohesion model indicated in Fig. 7 can be regarded as a pure linear birth-process. For simplicity by considering the evolution of a crack only, the relation for the transition probabilities will have the following form:

$$\frac{dP_{ij}(t)}{dt} = -\lambda_i P_{ij}(t) + \lambda_{j+1} P_{i,j+1}(t) \tag{4.9}$$

in which the transition intensities λ are one-step transitions from the state $i \to j$ and are functions of the involved energy release. The crack initiation, propagation and ultimate failure of polycrystalline solids such as metals can be studied in terms of such Markov processes. An important type of failure in structured solids is the fatigue failure, caused by the application of cyclic stresses. This has been investigated more recently in reference [109] on the basis of probabilistic principles. The formulation is carried out for the crack initiation process and crack propagation on the basis of the experimentally determined number of events in the initiation process and for the final event just before catastrophic failure of the material. Hence a pure birth discontinuous Markov process has been employed. The stochastic models of fatigue crack initiation and propagation and some experimental results are also given in references [105, 106].

So far as the behaviour of fibrous structures is concerned, a brief discussion on the breakdown of bonding, which leads to a complete failure of the microstructure is given below. The main interest here is the effect of bonding particularly of hydrogen-bonding (Fig. 5) when subjected to mechanical forces. As a consequence the unit-cell model indicated previously and the corresponding coincidence sites (hydroxyl groups (2, 2') in Fig. 5(d)) will experience a relative displacement, which is in the argument of the potential (4.4). This important parameter will be in general time-dependent and is characteristic for an individual bond. For the representation of the effect of bonding in a random network of fibers, it is more convenient, however, to introduce the number of bonds $^\alpha n(t)$. It is assumed that the initial number of bonds $^\alpha n_0$ at time $t = 0$ per elemental area of the bond ($^\alpha a$ in Fig. 5(b)) is experimentally accessible. Thus, in probabilistic terms, an initial set of events or observations can be specified. By the use of the general deformation theory, given subsequently in section 4.4 of this chapter, the occurring deformations can be represented by a Markov process.

Thus, by using again the number of bonds $^\alpha n(t)$, the characteristic differential equations for the bond deformation process is the Chapmann–Kolmogorov equation expressed in this case by:

$$\frac{dP_{ij}(t)}{dt} = \sum_k q_{ik} P_{kj}(^\alpha n(t)); \quad P_{ij}(0) = \delta_{ij} \tag{4.10}$$

4.3 Interaction effects

in which the transition probability $P_{ij}(^{\alpha}n(t))$ designates the transition of an individual bond from state i at time t to another state j at time $t + \Delta t$ in one step. The state i, in which a number of bonds are participating in the bonding of the elemental area $^{\alpha}a$, corresponds to an event i and, similarly, the state j corresponds to an adjacent event j. One may regard this representation as the probabilistic analogue to the deterministic rate equation reflecting the relative deformation of bonds with time. Alternatively, considering the probability distribution of $^{\alpha}n(t)$ at time t as a probability measure $\mathscr{P}_n(t)$, the Markov process related to the bond dissociation process can be seen as a simple death-process as follows:

$$\frac{d\mathscr{P}_n(t)}{dt} = -\lambda_n \mathscr{P}_n(t) + \lambda_{n+1} \mathscr{P}_{n+1}(t) \tag{4.11}$$

where the above relation is obtained by considering the rate of transition to depend on the state during the process. Thus, if the bond system is initially in the state i at time t and going in one step to the state j, ($j = i + 1$) at $t + \Delta t$, the transition probability can be designated by $\lambda_i \Delta t$, if the intensity is λ_i. The probability of remaining in the state i will then be $(1 - \lambda_i)\Delta t$ and the probability of changing to another state j will be $0(\Delta t)$.

In terms of these probability measures and assuming that the process is a linear one, the quantity $\lambda_n(t)$ can be replaced by $\lambda n(t)$ so that:

$$\frac{d\mathscr{P}_n(t)}{dt} = -\lambda n \mathscr{P}_n(t) + \lambda(n+1)\mathscr{P}_{n+1}(t); \quad 0 \leqslant n \leqslant n_0, \tag{4.12}$$

characterizing the evolution with time of the number of bonds in terms of the probability distribution, directly. The solution of the above differential equation is of the binomial form, i.e.:

$$\mathscr{P}_n(t) = \binom{n_0}{n} e^{-n_0 \lambda t} (e^{-\lambda t} - 1)^{n_0 - n}. \tag{4.13}$$

Considering the expected value of $n(t)$ or mean value and the variance, then:

$$E\{n(t)\} = \langle n(t) \rangle = n_0 e^{-\lambda t}, \quad \text{(a)}$$
$$\sigma^2(t) = n_0 e^{-\lambda t}(1 - e^{-\lambda t}). \quad \text{(b)} \tag{4.14}$$

Hence, for simplicity, if one considers the change of the number of bonds participating in the bond failure and by identifying $\langle ^{\alpha}n(t) \rangle \equiv n(t)$, relation (4.14(a)) gives in fact the solution of a deterministic rate equation, or:

$$\frac{dn(t)}{dt} = -\lambda n(t), \quad \text{where} \quad n(t)|_{t=0} = n_0. \tag{4.15}$$

Hence, it is seen that the simplest probabilistic models for the bond dissociation are indeed linear death processes.

However, it has been found from experimental investigations that the bond-breakage phenomenon in a fibrous structure occurs rather in a cooperative manner. This means that the individual bonds form a group during the dissociation process. For this purpose one can introduce a so-called cooperative index characteristic for a particular fibrous network. Since observations of fibrous networks indicate that the dissociation process occurs in a rather complex manner such that bonds can be formed and broken within the same mechanical state, a more general model is required. Thus, in general, a birth–death Markov process model can be used in the formulation of the bonding. In this context, it may be visualized that the breaking of bonds will occur in such a manner that energy is released, activating bond formation on other sites within the elemental area $^{\alpha}a$. In this case the transition intensity λ_n will depend not only on the number of bonds $^{\alpha}n(t)$, but also on the number of broken bonds, i.e. $(n_0 - n)$. The evolution of distributions of $^{\alpha}n(t)$ will then be given by the following form:

$$\left.\begin{aligned}\frac{d\mathscr{P}_{n_0}(t)}{dt} &= -\lambda n_0 \mathscr{P}_{n_0}(t), \\ \frac{d\mathscr{P}_n(t)}{dt} &= -\lambda n_0(n_0 - n + 1)\mathscr{P}_n(t) + \lambda(n+1)\mathscr{P}_{n+1}(t).\end{aligned}\right\} \quad (4.16)$$

The second equation in (4.16) may be solved with the initial condition:

$$\mathscr{P}_n(0) = \delta_{nn_0}$$

and the solution is then obtained as [100]:

$$\mathscr{P}_n(t) = \lim \frac{1}{2\pi i} \int_{\alpha-i\beta}^{\alpha+i\beta} \bar{\mathscr{P}}_n(s) e^{st} ds \quad (4.17)$$

where $\bar{\mathscr{P}}_n(s)$ is the Laplace transform of $\mathscr{P}_n(t)$ and can be expressed by:

$$\bar{\mathscr{P}}_n(s) = \frac{n_0!(n_0-n)!}{n!} \lambda^{(n_0-n)} \prod_{i=1}^{(n_0-n+1)} [s + i\lambda(n_0 - i + 1)]^{-1}. \quad (4.18)$$

Bailey [186] has investigated this form and has shown that for $n > n_0/2$ and an even n one can reduce (4.18) to:

$$\bar{\mathscr{P}}_n(s) = \sum_{i=1}^{(n_0-n+1)} \frac{\alpha n_i}{[s + i\lambda(n_0 - n + 1)]} \quad (4.19)$$

in which the coefficients αn_i take the following form:

$$\alpha n_i = \frac{(-1)^{i-1}(n_0 - 2i + 1)! n_0!(n_0 - n)!(n - i - 1)!}{n!(i-1)!(n_0 - i)!(n_0 - n - i - 1)!}. \quad (4.20)$$

Evaluating these coefficients and inverting $\bar{\mathscr{P}}_n(s)$ from equation (4.19) leads, then, to the distribution $\mathscr{P}_n(t)$ as well as to the mean value of $^\alpha n(t)$ for this type of process, as shown, for instance, by Haskey [110].

It is of interest to note that an energy release per bond designated by $\varepsilon(r)$ for a maximum cut-off distance r in the bond potential, can be estimated from experimental observation. This energy release is proportional to $\phi(r_0) - \phi(r_{\max})$, where $^{\alpha\beta}\phi(r)$ is the previously introduced interaction potential (Chapter 3). Thus, in a macroscopic sense, one can express the rate of total energy release to occur within a certain bonding area of the interface by:

$$\frac{dE(t)}{dt} = \varepsilon(r)\frac{dn(t)}{dt} = -\varepsilon\lambda(t)n(t). \tag{4.21}$$

Using the result of the simple Markov model, it is possible to interpret the transition intensity λ for the process by stating that:

$$E(t) = \varepsilon n(t)$$

and from (4.14(a)):

$$n(t) = n_0 e^{-\lambda t}, \quad \text{hence:} \quad E(t) = \varepsilon n_0 e^{-\lambda t}$$

so that the quantity

$$\lambda = \frac{1}{t}\ln\frac{\varepsilon n_0}{E(t)}. \tag{4.23}$$

It is to be noted that intensities λ are experimentally accessible, if the energy release averaged over the sample of the structure can be established by an appropriate experimental measurement.

4.4 GENERAL PROBABILISTIC DEFORMATION THEORY

In presenting a general probabilistic deformation theory of structured solids the notion of an abstract dynamical system will be again employed. This concept is equally useful in the analysis of the stability of microstructures during deformations. It is to be noted that in the present theory this concept is considered as more general in the sense of Loève [22], i.e. in the form of a triplet $[\mathscr{X}, \mathscr{F}, \mathscr{P}]$ and not with the restriction given in the definition by Rényi [17]. In the past, this concept has been used in a conventional manner in the analysis of elastic stability of continuous and discrete systems [112]. Hence, it will acquire in the present theory, a different meaning from that of the deterministic formulation.

(i) General kinematics

For the purpose of developing a probabilistic deformation theory, it is necessary to discuss briefly the kinematics and probabilistic structure of the deformation space. In continuum mechanics the deformation of a continuum is completely described by a set of vectors $\mathbf{u}(\mathbf{X}, t)$ as functions of the undeformed position vector $\mathbf{X}$ of a material point, and time t. Majuscules, generally, refer to the undeformed and miniscules to the deformed configuration of the material. In the probabilistic mechanics theory the formulation is distinctly different, since interaction effects are included. Here the deformation of a microelement is described generally by two random vectors $\mathbf{u}^i(\mathbf{X}, t)$ and $\mathbf{u}^s(\mathbf{X}, t)$, which are associated with the internal and surface deformations, respectively, i.e.:

$$\begin{aligned}\mathbf{u}^i(\mathbf{X}, t) &= \mathbf{x}(t) - \mathbf{X}, \\ \mathbf{u}^s(\mathbf{X}, t) &= \mathbf{g}(\mathbf{x}, t) - \mathbf{G}(\mathbf{X})\end{aligned} \quad (4.24)$$

in which $\mathbf{X}, \mathbf{G}$ are position vectors of any point inside and on the surface of a microelement with respect to an external coordinate frame (see Fig. 8). The vectors $\mathbf{x}, \mathbf{g}$ are their counterparts in the deformed configuration. Using a body frame in accordance with the given Definition 3 (Chapter 3), one can also write:

$$\mathbf{X} = \mathbf{O} \cdot \mathbf{Y} + \mathbf{R} \quad (4.25)$$

where $\mathbf{O}$ is a random orientation matrix and $\mathbf{R}$ the random position vector to the center of mass of the microelement. Thus relations (4.24) can be rewritten as follows:

$$\mathbf{u}^i(\mathbf{Y}, t) = \mathbf{o} \cdot \mathbf{y}(t) - \mathbf{O} \cdot \mathbf{Y} + \mathbf{r}(t) - \mathbf{R} \quad (4.26)$$

and by using a moving surface coordinate frame $\mathbf{H} \colon \mathbf{G} = \mathbf{O} \cdot \mathbf{H} + \mathbf{R}$

$$\mathbf{u}^s(\mathbf{H}, t) = \mathbf{o} \cdot \mathbf{h}(t) - \mathbf{O} \cdot \mathbf{H} + \mathbf{r}(t) - \mathbf{R}. \quad (4.27)$$

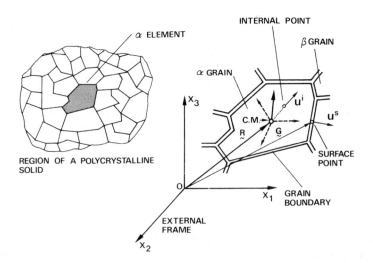

FIG. 8. Kinematics of an element (undeformed configuration).

4.4 General probabilistic deformation theory

Hence one can introduce the generalized deformation $^{\alpha}\hat{\mathbf{u}}$ of a microelement as a function of $\mathbf{u}^i$ and $\mathbf{u}^s$ such that:

$$^{\alpha}\hat{\mathbf{u}}(\mathbf{X}_{CM}, t) = f(^{\alpha}\mathbf{u}(\mathbf{X},t)) = f(\mathbf{u}^i(\mathbf{X},t) \oplus \mathbf{u}^s(\mathbf{X},t)) \quad (4.28)$$

or equivalently in terms of Y, H and t

$$^{\alpha}\hat{\mathbf{u}}(^{\alpha}\mathbf{X}_{CM}, t) = f(\mathbf{u}^i(\mathbf{Y},t) \oplus \mathbf{u}^s(\mathbf{H},t)) \quad (4.29)$$

It is to be noted that $\mathbf{u}^i(\mathbf{X}, t)$ and $\mathbf{u}^s(\mathbf{X}, t)$ are not continuous functions in the strict sense, but may be regarded in general to be at least piecewise continuous and that derivatives of all orders within some compact support exist. In this sense the generalized deformation $^{\alpha}\hat{\mathbf{u}}(\mathbf{X}, t)$ will be a discrete random function $^{\alpha}\mathbf{u}(^{\alpha}\mathbf{X}, t)$, $\alpha \in M$ in the meso-domain M. From a systems theory point of view, the generalized microdeformations have been given in previous work [36, 89] in the form of:

$$^{\alpha}\hat{\mathbf{u}}(^{\alpha}\mathbf{X}_{CM}, t) = \langle L_1, \mathbf{u}^i(\mathbf{X},t) \rangle \oplus \langle L_2, \mathbf{u}^s(\mathbf{X},t) \rangle$$
$$= \hat{\mathbf{u}}^i(^{\alpha}\mathbf{X}_{CM}, t) \oplus \hat{\mathbf{u}}^s(^{\alpha}\mathbf{X}_{CM}, t) \quad (4.30)$$

where L_1, L_2 are continuous operators within the compact support in which $\mathbf{u}^i$, $\mathbf{u}^s$ are defined, respectively. It is considered in the present theory that a function of the form (4.30) can always be established, although the operators L_1, L_2 may not be simple. A more detailed discussion on certain forms of the operators is given in references [36, 101, 103]. Since by definition the basic quantities are random variables, the above deformations are also random variables and during a more general deformation become random processes.

Analogous kinematic relations will apply to fibrous structures and other discrete structures of solids. It is important to recognize, that material points inside an element α and points in the interaction zone (grain boundary in polycrystalline solids) must be distinguished with respect to the internal and external coordinate frames, respectively. A more detailed discussion on the two groups of structured solids considered here has been given in references [36, 96].

(ii) *General probabilistic deformation theory*

The general kinematics of structured solids have been considered in the foregoing section. It is intended to develop a probabilistic deformation theory applicable to such solids in general. For this purpose it may be recalled from the discussion of the mechanical state of a microelement and the probabilistic structure of the state space $S = \mathscr{X}$ (Chapter 3), that it is convenient to use the subspace $\mathscr{U} \subset \mathscr{X}$ in the formulation of the deformation theory. It has also been shown that the state vector $^{\alpha}\mathbf{s}$ contains several components each of which belong to a subspace of $\mathscr{X}$. Thus considering deformations only, it is assumed that the outcomes fall within a certain window set (see remark after Def. 11).

96 Probabilistic Mechanics of Solids

The notion of generalized microdeformations $\hat{\mathbf{u}}$ like those of generalized microstresses $\hat{\sigma}$ will be used from now on. It amounts to a discretisation of these quantities in the corresponding domains of the material body.

It is evident from the foregoing discussion on the kinematics of solids, that in the case of deformations only, one has:

$$^{\alpha}\mathbf{u}^i \in U^i, \quad ^{\alpha}\mathbf{u}^s \in U^s, \quad U^i \oplus U^s = \mathcal{U} \subset \mathcal{X} \qquad (4.31)$$

where U^i, U^s designate the internal and surface deformation space, respectively, and where both are contained in $\mathcal{X}$. Since $\mathcal{X}$ has been shown to be a measurable space, the deformation space $\mathcal{U}$ as well as the subspaces U^i, U^s are also measurable. Thus a measure $\mathscr{P}^{\mathbf{u}}$ on $\mathcal{U}$ can be introduced with the following properties:

$$\left.\begin{aligned}
&\text{(i)} \quad 0 \leqslant \mathscr{P}^{\mathbf{u}}\{E_n\} \leqslant 1 \quad \text{for} \quad \forall E_n \in \mathscr{F}^{\mathbf{u}}, \quad n = 1, 2, \ldots, \\
&\text{(ii)} \quad \mathscr{P}^{\mathbf{u}}\{E_1 \cup E_2\} = \mathscr{P}^{\mathbf{u}}\{E_1\} + \mathscr{P}^{\mathbf{u}}\{E_2\} - \mathscr{P}^{\mathbf{u}}\{E_1 \cap E\} \\
&\qquad \text{if } E_1, E_2 \in \mathscr{F}^{\mathbf{u}} \text{ and } E_1 \cap E_2 \neq \phi, \\
&\text{(iii)} \quad \mathscr{P}^{\mathbf{u}}\left\{\bigcup_n E\right\} = \sum_n \mathscr{P}^{\mathbf{u}}\{E_n\}, \quad \text{if } E_n \cap E_m = \phi, n \neq m \\
&\qquad n, m = 1, 2, \ldots, \\
&\text{(iv)} \quad \mathscr{P}^{\mathbf{u}}\{\mathcal{U}\} = 1.
\end{aligned}\right\} \qquad (4.32)$$

This measure is closely related to the distribution function in probability theory and is called the distribution function of the microdeformations $^{\alpha}\mathbf{u} \equiv {^{\alpha}\hat{\mathbf{u}}}$. Hence the corresponding probability space can be represented by $[\mathcal{U}, \mathscr{F}^{\mathbf{u}}, \mathscr{P}^{\mathbf{u}}]$. It has been shown in ref. [115] that $\mathscr{P}^{\mathbf{u}}$ satisfied the condition of regularity. As a consequence, one can define a random vector $\mathbf{u} \in \mathcal{U}$ as a $\mathscr{P}^{\mathbf{u}}$-regular measurable function in $\mathcal{U}$ (see also the definitions given by Doob [33] and Billingsley [113]).

The identification of $[\mathcal{U}, \mathscr{F}^{\mathbf{u}}, \mathscr{P}^{\mathbf{u}}]$ as the space of all $\mathscr{P}^{\mathbf{u}}$-regular measurable functions of $\mathbf{u}$ with a probabilistic function space forms the basis for the mathematical structure of the general deformation theory of discrete solids. Moreover, this leads directly by recognizing the duality with the corresponding stress space (see also ref. [35]) to the establishment of the response behaviour of discrete solids in an operational form. In this context the general state space $\mathcal{X}$ may be thought of as being composed of the deformation space $\mathcal{U}$ and the stress space Σ. In terms of system theory, the input to a microelement could be regarded as an element of $\mathcal{U}$ and the output as an element of Σ. For a given physical system a mapping between $\mathcal{U}$ and Σ will always exist, whereby the topological structure of one of the spaces, say Σ, may be given in terms of a non-degenerate bilinear form with respect to this mapping (see also Moreau [87] and Tonti [114]). This, however, involves the concept of a material operator as set down by Postulate 4. This will not be pursued here, but discussed further in the following subsection.

4.4 General probabilistic deformation theory

The function space approach requires the use of a set of norms or semi-norms in the associated probability space. Thus considering $\mathcal{U}$ as the space of all $\mathscr{P}^u$-regular measurable and bounded functions of **u** which are discrete, one can define the expected value of **u** as follows:

$$E\{\mathbf{u}\} = \Sigma \mathbf{u} p\{E(\mathbf{u})\} = \Sigma \mathbf{u} \Delta \mathscr{P}^u \qquad (4.33)$$

which is also the mean value of **u** to be denoted subsequently by $\langle \mathbf{u} \rangle$. If in an experiment the expected value of **u**, which is representative of a macrodeformation, i.e. $\langle \mathbf{u} \rangle$ is equal to zero, although microdeformations may still exist, it is seen that $|\langle \mathbf{u} \rangle|$ or $|E\{\mathbf{u}\}|$ is in fact a semi-norm (Yosida [7]). In this case $\mathcal{U}$ will be a Fréchet space with the semi-norm stated above. If $\mathcal{U}$ is a linear topological vector space, it can be suitably topologized by a family of semi-norms and is then a locally convex space. Since the properties of a Fréchet space closely resemble those of a Banach space, the analysis in either of these spaces is similar [3, 116, 117]. Thus, if in an experiment, one is concerned with finding the average value of the magnitude of **u**, a similar definition to (4.33) will satisfy the properties of a norm and therefore $\mathcal{U}$ becomes a Banach space. The standard deviation of the discrete microdeformations can be expressed in terms of the measure by:

$$D^u = \{\Sigma |\mathbf{u}_n - \langle \mathbf{u}_n \rangle|^2 \Delta \mathscr{P}^u\}^{\frac{1}{2}} \qquad (4.34)$$

where $|D^u|$ satisfies the properties of a semi-norm and $\mathcal{U}$ is a Fréchet space X. Evidently, if higher-order statistics of a random variable or random function are used other definitions of norms or semi-norms can be given.

The deformation kinematics of discrete media have been briefly considered at the beginning of this section. The evolution of the deformations is, however, of interest in a general deformation process. For this purpose one can consider a sequence of deformations $\mathbf{u}(\mathbf{X})$ indexed by the time t and designated by $\mathbf{u}_t(\mathbf{X})$ as a random process (see, for instance, Yaglom [30]). Thus $\mathbf{u}_t(\mathbf{X})$ defines a random deformation process in the deformation space where t belongs to the positive half of the real line $\mathbb{R}^+$. The function $\mathbf{u}(\mathbf{X}, t)$ is a random function that for a fixed time $t \in \mathbb{R}^+$ is a random vector $\mathbf{u}(\mathbf{X})$ in $[\mathcal{U}, \mathscr{F}^u, \mathscr{P}^u]$.

Considering the probability function space $[\mathcal{U}, \mathscr{F}^u, \mathscr{P}^u]$ for any particular time t, the whole deformation process will then be represented by a set of function spaces leading to the notion of a product space [11, 116, 118]. In particular, if $\mathbb{R}^+ = [0, \infty)$, where for each $t_r \in \mathbb{R}^+$, $r = 1, 2, \ldots, N$ there corresponds a triplet $[\mathcal{U}, \mathscr{F}^u, \mathscr{P}^u]$ an N-fold product of these spaces leads to a product space in which $\mathbf{u}(\mathbf{X}, t)$ becomes a measurable function. For convenience the N-fold product space can be extended to infinity and designated by $\mathcal{U}_\infty, \mathscr{F}_\infty, \mathscr{P}^u$ so that $\mathbf{u}(\mathbf{X}, t)$ may be regarded as a t-continuous random function. Alternatively, one can consider $[\mathcal{U}, \mathscr{F}^u, \mathscr{P}^u]$ and a one-parameter family $\{L_t\}$ of transformations L_t such that:

$$L_t: \mathcal{U} \to \mathcal{U} \quad \text{for} \quad \forall t \in \mathbb{R}^+ = [0, \infty); \quad \mathbf{u}_t(\mathbf{X}) \in \mathcal{U}. \qquad (4.35)$$

In this sense the deformation process is defined as a measurable function, i.e. a random function which is generated by the random endomorphism $\{L_t\}$ for $\forall\, t \in \mathbb{R}^+$ (see remark in section 2.2 (ii) of Chapter 2). The σ-algebra $\mathscr{F}_\infty^u$ of this product space has a countably finite number of Borel sets $E_1(\mathbf{u})$, $E_2(\mathbf{u}), \ldots, E_N(\mathbf{u})$ corresponding to the sequence $t_1, t_2, \ldots, t_N \in \mathbb{R}^+$ and where each Borel set is obtained from the other by the automorphism or its inverse. If $t_1 < t_2 < \ldots < t_N$, then:

$$L_{\Delta t_r} E_r(\mathbf{u}) = E_{r+1}(\mathbf{u}); \quad r = 1, 2, \ldots, N-1; \quad \Delta t_r = = t_{r+1} - t_r. \quad (4.36)$$

It can be shown that if $\mathscr{P}^u$ is a regular measure on E_r and L_t is defined so as to satisfy eqn (4.36), E_{r+1} is also $\mathscr{P}^u$-regular measurable. One can generalize this statement to any Borel sets $E_r(\mathbf{u})$; $r = 1, 2, \ldots, N$ in $\mathscr{F}_\infty$. Thus, it may be concluded that at any time during a deformation process the random deformations are $\mathscr{P}^u$-regular measurable.

In particular, in the case of purely elastic deformations under a steady load, one can write that:

$$\mathscr{P}^u\{E_{r+1}(\mathbf{u})\} = \mathscr{P}^u\{E_r(\mathbf{u})\}; \quad r = 1, 2, \ldots, N-1 \quad (4.37)$$

in which the probability measure is independent of time. This means that the deformation process is a strictly stationary random process (see also Bochner [119], Doob [33] and others). The inelastic material behaviour involves, however, the change of the probability distribution of microdeformations with time. From a measure theoretical point of view, if the Borel set $E_r \in \mathscr{F}_\infty$ at time $t_r \in \mathbb{R}^+$ is given, the probability measure on the set $E_s \in \mathscr{F}_\infty$ at time $t_s \in \mathbb{R}^+$ must be established. This can be done by using a conditional probability measure introduced by Kolmogorov [1] and Rényi [17]. It is defined as an extended real-valued set function on $\mathscr{F}_\infty$ with properties analogous to those specified in relations (i–iv) in (4.32). Using the concept of conditional probability, it is evident that to each automorphism L_{t_r} there corresponds a conditional probability measure $\mathscr{P}\{E_{r+1}|E_r\}$ such that whenever:

$$\begin{aligned} L_t E_r(\mathbf{u}) &= E_{r+1}(\mathbf{u}), \\ \mathscr{P}^u\{E_{r+1}\} &= \mathscr{P}^u\{E_{r+1}|E_r\}\,\mathscr{P}^u\{E_r\}. \end{aligned} \quad (4.38)$$

The above relation between the measures $\mathscr{P}^u\{E_{r+1}\}$ and $\mathscr{P}^u\{E_r\}$ can be obtained by considering two Borel sets E_n, E_s for which:

$$\left.\begin{aligned} E_n \cap E_s &= E_s, \text{ if } E_n \supset E_s \\ \text{and} \quad \mathscr{P}^u\{E_n \cap E_s\} &= \mathscr{P}^u\{E_s\}. \end{aligned}\right\} \quad (4.39)$$

By inserting $r+1$ for s, it is seen that:

$$\mathscr{P}^u\{E_r \cap E_{r+1}\} = \mathscr{P}^u\{E_{r+1}\}.$$

4.4 General probabilistic deformation theory

Using Kolmogorov's definition of the conditional probability shows that:

$$\left. \begin{array}{r} \mathscr{P}^u\{E_{r+1}|E_r\} = \dfrac{\mathscr{P}^u\{E_{r+1} \cap E_r\}}{\mathscr{P}^u\{E_r\}} \\ \text{and} \qquad \mathscr{P}^u\{E_{r+1}\} = \mathscr{P}^u\{E_{r+1}|E_r\}\mathscr{P}^u\{E_r\} \equiv (4.38). \end{array} \right\} \quad (4.40)$$

This relation is readily generalized to read:

$$\mathscr{P}^u\{E_n\} = \mathscr{P}^u\{E_1\} \prod_{r=1}^{n-1} \mathscr{P}^u\{E_{r+1}|E_r\} \quad (4.41)$$

which is valid for any sequence $E_1, E_2, \ldots, E_r, \ldots, E_N$ corresponding to the time sequence $t_1 < t_2 < \ldots < t_r < \ldots < t_N$ and a set of conditional probabilities $\mathscr{P}^u\{E_{r+1}|E_r\}$, $r = 1, 2, \ldots, N-1$. The relations (4.40, 4.41) are of utmost significance in the analysis of the deformation process since they permit the determination of probability distribution of deformations at any time during the process, if the distribution at any other time or the initial one is known. This is easily recognized when eqn. (4.38) is rewritten in a more explicit form, i.e.:

$$\mathscr{P}^u\{E_{r+1}, t_{r+1}\} = \mathscr{P}^u\{E_{r+1}, t_{r+1}|E_r, t_r\}\mathscr{P}^u\{E_r, t_r\} \quad (4.42)$$

in which E_r corresponds to $t_r \in \mathbb{R}^+$ and E_{r+1} to $t_{r+1} > t_r$. It is apparent that for a deformation process in which $\mathscr{P}^u\{E_{r+1}, t_{r+1}|E_r, t_r\}$ depends on the time difference Δt_r only (eqn. (4.36)), the sequence of deformations $\mathbf{u}_t(\mathbf{X}) \in [\mathscr{U}, \mathscr{F}^u, \mathscr{P}^u]$ together with the conditional probability measure will describe generally a homogeneous stochastic process. Letting for simplicity $t_{r+1} = t$ and $t_r = 0$ eqn. (4.42) becomes:

$$\mathscr{P}^u\{t\} = \mathscr{P}^u\{\mathbf{u}(t)\} = \mathscr{P}(t)\mathscr{P}^u(0) = \mathscr{P}(t)\mathscr{P}^u\{\mathbf{u}(0)\} \quad (4.43)$$

indicating the possibility of evaluating the distribution of microdeformations at time t from the knowledge of the initial distribution $\mathscr{P}^u(0)$. It has been shown in previous work [35, 36] that a particular stochastic process, i.e. the Markov process, can be used to characterize the deformational behaviour of discrete media. To this effect considering the Banach space $X(\mathscr{U})$ of all measures $\mathscr{P}^u$, then the conditional probability $\mathscr{P}\{t_r, t_{r+1}\}$ can be shown to be a contraction operator on $X(\mathscr{U})$ to $X(\mathscr{U})$, i.e. $\|P(E)\| \leq 1$. In the theory of Markov processes this operator is known generally as transition probability. For instance, considering a closed time interval $[t, s] \in \mathbb{R}^+$, subdividing it into smaller intervals and selecting a point $\tau > t, \tau \in [t, s]$, the transition probability will satisfy the well-known Chapman–Kolmogorov functional relation, i.e.:

$$P\{t, s\} = \int_{\mathscr{U}} P\{t, \tau\} \, dP\{\tau, s\}. \quad (4.44)$$

For the homogeneous type process this relation due to the dependence of P on the time difference $\{t-s\}$ only, becomes then:

$$P\{t-s\} = \int_{\mathcal{U}} P\{t-\tau\}\,dP\{\tau-s\}. \qquad (4.45)$$

The above relations (4.44, 4.45) are in the general deformation theory of discrete solids of utmost importance since they connect the stochastic theory with the functional analytic formulation presented in this analysis.

The material generally can pass through a countable number of states during a deformation. A Borel set related to the deformation space at time t_r corresponds then to a state i and the Borel set at time t_s to another state j. It is evident that for fixed i, j eqn. (4.44) can be expressed as follows:

$$P_{ij}(t_r + t_s) = \sum_k P_{ik}(t_r) P_{kj}(t_s) \qquad (4.46)$$

for all possible states i, j. The quantity on the left side of (4.46) is an element of the matrix $\underline{P}(t_r + t_s)$ and hence relation (4.44) rewritten in matrix form becomes:

$$\underline{P}(t_r + t_s) = \underline{P}(t_r)\underline{P}(t_s) \qquad (4.47)$$

displaying the semi-group property. It has been shown by Kampé de Fériet [62], that from a statistical mechanics point of view the semi-group property is related to the irreversible behaviour of the system. In the present context this property can be used to establish criteria for the micro-structural stability of discrete solids as discussed subsequently. Following Doob[33] and Kendall [120], the elements $P_{ij}(t)$ of the transition matrix in a time-homogeneous process representing the transition from a state i to a state j during the deformation of the material can be shown to satisfy the limiting conditions given below. Thus for $t, s \in \mathbb{R}^+$, $P_{ij}(t)$ of such a process, is a uniformly continuous function of time (see also Bharucha-Reid [121]) so that:

$$|P_{ij}(t+s) - P_{ij}(s)| \leq 1 - P_{ij}(t) \qquad (4.48)$$

and where the limits

$$\left.\begin{array}{c} \displaystyle\lim_{t \to 0^+} \frac{1 - P_{ii}(t)}{t} = -q_{ii} < \infty, \\[2ex] \displaystyle\lim_{t \to 0^+} \frac{P_{ij}(t)}{t} = q_{ij} \end{array}\right\} \qquad (4.49)$$

exist and are measurable for all $t \in \mathbb{R}^+$ so that

$$1 - P_{ij}(t) \leq 1 - e^{-q_{ij} \cdot t}. \qquad (4.50)$$

4.4 General probabilistic deformation theory

In the above expressions $q_{ij} \in \underline{Q}$ is a measure of the relative transition and hence $\underline{Q}$ is also called the transition intensity matrix. Using the statement in (4.50) the inequality (4.48) can be expressed by:

$$|P_{ij}(t+s) - P_{ij}(s)| \leqslant 1 - e^{-q_{ij} \cdot t}$$

and hence for any $P(t)$:

$$\lim_{t \to 0} |\{\underline{P}(t+s) - \underline{P}(s)\} P(t)| = 0 \quad \text{for} \quad \forall s, \geqslant 0 \qquad (4.51)$$

or equivalently

$$\lim_{t \to s} [\underline{P}(t) - \underline{P}(s)] P(t) = 0 \quad \text{for} \quad \forall s, \geqslant 0. \qquad (4.52)$$

In view of these properties of the transition probability the time-homogeneous or steady-state deformation process can be rigorously defined in the following manner:

$$\left.\begin{array}{l} \text{(i) } \underline{P}(t) \text{ is a contraction on the space } X(\mathcal{U}) \to X(\mathcal{U}), \|\underline{P}(t)\| \leqslant 1; \\ \text{(ii) } \forall s, t \in \mathbb{R}^+, s, t \geqslant 0 \quad \text{and} \quad E \in \mathscr{F}^u \text{ of } \mathcal{U} \text{ correspond to } t \\ \quad \lim_{t \to s} [\underline{P}(t) - \underline{P}(s)] \mathscr{P}\{E\} = 0; \\ \text{(iii) } \forall s, t \in \mathbb{R}^+ \\ \quad \underline{P}(t+s) = \underline{P}(t)\underline{P}(s) \quad \text{with} \quad \underline{P}(0) = \underline{I} \text{ (identity matrix).} \end{array}\right\} \qquad (4.53)$$

These properties are not always satisfied since there are other stages in a general deformation process than those described above.

In general, however, for each stage the Chapman–Kolmogorov relation (4.44) will hold, although the semi-group property displayed by the transition probabilities may not be satisfied. Relation (4.44) in the form of a matrix differential equation can be written as:

$$\frac{d\underline{P}(t,s)}{dt} = \underline{Q}(t)\underline{P}(t,s); \quad \underline{P}(0) = \underline{I}. \qquad (4.54)$$

This relation is a key equation in the functional analytic formulation of deformations of discrete solids. It represents an equation of evolution the solution of which will be further discussed below. In closing this section, it should be noted that the probabilistic and topological structure of deformations have been given without reference to the topology and measures on the associated stress-space. However, since there exists an intrinsic relation between these spaces, measures chosen *a priori* on the deformation space will restrict the choice of measures on the stress-space. This is due to the required mapping between the spaces for which a measurable transformation and the property of invertibility of the material operator (Postulate P.4) is needed.

4.5 STABILITY OF MICROSTRUCTURES

In continuation of the above discussion of the probabilistic deformation theory, the various stages that may occur during a general deformation process are considered now. In particular the significance of the transition probabilities with regard to a stable or unstable deformation process will be investigated.

In continuum physics the behaviour of conservative systems is usually represented by hyperbolic type differential equations which are associated with group transformations T_t on $\mathscr{X}$ with a domain $-\infty < t < \infty$. Dissipative systems, however, correspond to parabolic type differential equations that induce only semi-groups $\{T_t\}$ defined for $t > 0$. The elastic stability of dynamical systems both for the continuous and discrete case have been investigated in a comprehensive study by Knops and Wilkes [112] (see also [122, 123]). It has been shown by Kolmogorov [1] that under fairly broad assumptions, the transition probabilities defined earlier can be obtained from certain parabolic differential equations, which are concerned with the evolution of field variables of a physical system. The problem of integration of such equations has been widely investigated, mostly in terms of an operator formalism [124–127]. Thus Yosida's treatment [7, 128] considers a field variable as an element of the Banach space $\mathscr{X}$ and depending on the real parameter t. The solution of the corresponding evolution equation is then given in terms of an operator $A(t)$ which generally is unbounded and has a domain $\mathscr{D}(A(t))$ and a range $\mathscr{R}(A(t))$ both in $\mathscr{X}$. Similar considerations of equations of evolution and their solutions are due to Mizohata [129] in which field variables may have arbitrary initial values, but where the operator $A(t)$ is bounded. Both the cases of parabolic and hyperbolic systems are considered. From the properties of the transition probabilities $P(t)$ and the matrix differential equation (4.54), it follows that in the case of elastic deformations only, the stochastic deformation process is time-independent and corresponds to a stationary random process. The elastic deformations under constant load application belong to a subset of the more general product space $[\mathscr{U}_\infty, \mathscr{F}_\infty, \mathscr{P}^u]$ on which the measures generated by the transition probabilities (4.38) are such that:

$$|\mathscr{P}^u\{t+s\} - \mathscr{P}^u\{t\}| \to 0 \quad \text{for} \quad \forall s, t \in \mathbb{R}^+ \quad (4.55)$$

which implies

$$\mathscr{P}^u\{t+s\} \stackrel{\text{a.e.}}{=} \mathscr{P}^u\{t\} \quad (4.56)$$

and where $\mathscr{P}^u\{E\}$ is time-independent. The following properties of the transition probabilities will hold:

(i) the transition probability $\mathscr{P}^u(t, t+\Delta t)$ of the process $\mathbf{u}(t)$ changing from state i to another state j in the interval Δt is zero.
(ii) $\mathscr{P}^u(t, t+\Delta t)$ of no change in the interval Δt is equal to 1.

4.5 Stability of microstructures

In view of the above statements the Kolmogorov relation (4.54) reduces to:

$$\frac{d\underline{P}^u}{dt} = 0 \quad \text{since} \quad \underline{P}^u = \text{constant matrix;} \quad \underline{Q} = 0 \qquad (4.57)$$

which indicates that the elements $q_{ij}(t) \in Q(t)$ are time-independent infinitesimal constants, i.e.:

$$q_{ii} = 0, \quad \text{for} \quad i = 0, 1, \ldots; \quad q_{ij} = 0 \quad \text{for} \quad j = i+1. \qquad (4.58)$$

Equation (4.57) can be solved for an initial value $\underline{P}(0) = \underline{I}$ giving

$$\underline{P}^u(t) = \underline{I} \quad \text{or} \quad P^u_{ij}(t) = \delta_{ij}; \quad 0 < t \leqslant t_1. \qquad (4.59)$$

This result suggests that the probability distribution of the stochastic microdeformations within the reversible or elastic response of the discrete material under a constant load applied within $[0, t]$, remains constant with time. It is consistent with that discussed below concerning the transient and steady-state response. Considering the latter first, the properties of the transition probabilities have been rigorously defined in (4.53). It is of interest to employ the transition probabilities and the corresponding transition intensity matrix to represent the evolution of the distributions $\mathscr{P}\{\mathbf{u}(t)\}$. It may be noted that the previous designation of the stochastic deformation process has been changed to $\mathbf{u}(t)$. During the steady-state deformation of a discrete material the random vector $\mathbf{u}(t)$ changes from one state to another with a certain intensity λ, that is directly related to the elements of the transition matrix. The significance of λ with respect to the microstructural stability during deformations will be shown subsequently. The states a material can pass through are again identified by the index of the event set or the Borel set to which $\mathbf{u}(t)$ belongs at the corresponding instant of time. Thus, if within a small time interval Δt, $\mathbf{u}(t)$ changes from an initial state 0 to a state 3 for instance, the intensity of this change will be given precisely by the element q_{03} of the matrix $\underline{Q}$. It is assumed, however, that Δt is small enough so that a change of $\mathbf{u}(t)$ is permitted from state i to an adjacent state j only, when $j = i+1$ and $i = 0, 1, 2, \ldots$. Hence the following characteristics of a steady-state deformation process may be stated:

(i) The intensity with which the random deformation $\mathbf{u}(t)$ changes from a state i to its adjacent state j in the interval Δt is $\lambda \Delta t + 0(\Delta t)$, where λ is a positive constant and $0(\Delta t)$ is the order of magnitude of Δt.

(ii) The intensity with which $\mathbf{u}(t)$ does not change from a state i to a state j in Δt is $(1 - \lambda \Delta t - 0(\Delta t))$.

(iii) The intensity with which $\mathbf{u}(t)$ changes from a state i to any other state except j, $(j = i+1)$ in the interval Δt is $0(\Delta t)$.

From the above criteria, it is apparent that the elements of the intensity matrix for a steady-state deformation process will be:

$$\left.\begin{array}{ll} q_{ii} = -\lambda & \text{for } i = 0, 1, \ldots, \\ q_{ij} = \lambda & \text{for } j = i+1, \\ \phantom{q_{ij}} = 0 & \text{otherwise.} \end{array}\right\} \quad (4.60)$$

Hence, the form of the intensity matrix $\underline{Q}$ shows that the steady-state deformation process can be represented by a Poisson process. A more detailed account of such processes is given amongst others by Rényi [111], Prékopa [130] and Urbanik [131].

The notion of the intensity factor λ introduced above is rather important in the present theory since it indicates the degree of transition from one state to another and hence, whether the microstructure of the material remains stable or not during a particular stage of deformation, in this case under the steady-state and isothermal conditions. Some further remarks on this parameter should be made here. In the theory of Markov processes, if the initial state in a time homogeneous process is known, the change from this state may occur at any random instant of time. The time elapsed until $u(t)$ enters a new state is referred to as the waiting time τ. It is easily shown that the random variable τ has an exponential distribution with the parameter λ as a non-negative constant, i.e. $F(\tau) = e^{-\lambda \tau}$. The value of $\lambda = \infty$ is not excluded. If the intensity $\lambda = 0$, the process $\mathbf{u}(t)$ remains in the same state. If, however, $\lambda = \infty$, the process will leave the state instantaneously (absorbing state). For any intermediate value of λ, the transition intensity q_{ij} will be:

$$\left.\begin{array}{ll} 0 \leqslant q_{ij} < \infty & \text{for } i \neq j, \\ q_{ij} = q_{ii} = -q_i & \text{for } i = j, \end{array}\right\} \quad (4.61)$$

which implies, in accordance with (4.60), that for $0 \leqslant \lambda < \infty$ any state i will be a stable one. Once the system is in a stable state, it remains there with probability one for a positive period of time. If the system has no instantaneous states, i.e. all transition intensities are finite, a necessary and sufficient condition for a stable transition mechanism to occur is given by:

$$\sum_{n=1}^{\infty} \frac{1}{\lambda_{\mathbf{u}(\tau_n)}} = \infty \text{ with probability } 1 \quad (4.62)$$

in which τ_n is the time instant of leaving the state $i_n = u_n(\tau_{n-1})$. From the above derived criteria for the intensity, the Kolmogorov differential equation for the steady state can be written as:

$$\frac{dP_{ij}^{\mathbf{u}}(t)}{dt} = -\lambda P_{ij}^{\mathbf{u}}(t) + \lambda P_{i,j-1}^{\mathbf{u}}(t); \quad t_2 \leqslant t < \infty \quad (4.63)$$

where t_2 is the instant when the steady state begins. In general, this time will coincide with the time instant of the upper limit for the purely elastic material

4.5 Stability of microstructures 105

behaviour. Under circumstances such as creep deformations for instance, t_2 will refer to the upper limit of the transient stage occurring within a fixed time interval $[t_1, t_2]$. A combination of the reversible and steady-state stages in terms of the intensities λ in (4.58) and (4.60) together with equation (4.54) forms an equivalent representation to the phenomenological model of the material behaviour known as the elastic plastic response under a constant load. This is indicated in Fig. 9 below. It is seen that Fig. 9(a) corresponds to the perfectly plastic response, whilst (b) represents an elastic–perfectly plastic solid and (c) the case of an elastic–plastic with strain-hardening material. It may be recognized that in the phenomenological formulation of the material response, the change of the material from a reversible to an irreversible state is only represented by a single point. In the actual material behaviour this change must certainly occur within a finite period of time. Hence the present analysis allowing for a transient to occur is more in line with a realistic representation of the material behaviour. This transient in the probabilistic formulation is seen as a rearrangement of the microstructure. The subsequent discussion refers in particular to the response due to a constant load application with time (Fig. 9(d)). Thus, for the conditions of creep, however, the transient stage will

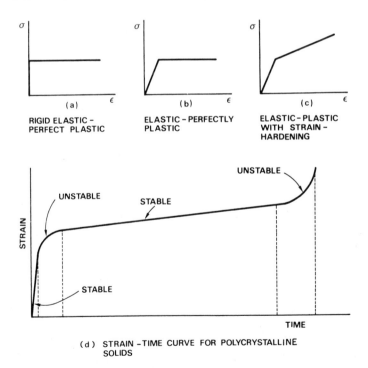

FIG. 9. Phenomenological models of the elastic–plastic response and time-dependent behaviour of polycrystalline solids.

be generally of a longer duration as discussed below. Using an iteration scheme [115], the general solution of equation (4.63) is obtained as:

$$\left.\begin{aligned} P_{ij}(t) &= \frac{[\lambda(t-t_2)]^{(j-i+1)}}{(j-i+1)!} e^{-\lambda(t-t_2)} \quad \text{for} \quad j \geqslant i, t_2 \leqslant t < \infty \\ &= 0 \quad \text{for} \quad j < i. \end{aligned}\right\} \quad (4.64)$$

For $j \geqslant 1$ one gets:

$$\left.\begin{aligned} 0 &\leqslant e^{-\lambda(t-t_2)} \leqslant 1; \quad t_2 \leqslant t < \infty \quad &\text{(a)} \\ \text{and} \quad P_{ij}(t) &= \frac{[\lambda(t-t_2)]^{j-i}}{(j-i)!} \geqslant 0; \quad t_2 \leqslant t < \infty \quad &\text{(b)} \end{aligned}\right\} \quad (4.65)$$

so that $P_{ij}(t)$ is always $\geqslant 0$ for $j \geqslant i$ and $P_{ij}(t) = 0$ for $j < i$. Hence $P_{ij}(t) \geqslant 0$ for all i,j and is, as shown previously in (4.53) the contraction operator, i.e. $\|\underline{P}(t)\| \leqslant 1$. Taking the initial conditions to be $P_{00}(t_2) = 0, P_{01}(t_2) = 1$, i.e. at the beginning of the steady state at t_2 will give by using (4.63, 4.64) a transition probability as follows:

$$P_{i,i+1}(t) = e^{-\lambda(t-t_2)}; \quad j = i+1, \quad t_2 \leqslant t < \infty \quad (4.66)$$

in which $P_{i,i+1}(t)$ is known as the one-step transition probability.

Employing this transition probability permits the assessment of the distribution of microdeformations during the steady state from two successive experimental observations since

$$\mathscr{P}^u(t) = e^{-\lambda(t-t_2)} \mathscr{P}^u(t_2) \quad (4.67)$$

according to the relations between $\mathscr{P}^u(t)$ and $\mathscr{P}^u(t_2)$ given in eqn. (4.43). During the steady-state stage the deformations are stable and the intensity satisfies condition (4.65a). This stage remains stable so long as $\lambda \equiv \lambda_s$ (steady state) is smaller than unity.

From the point of view of microstructural stability, the transient stage is the most significant. This stage as mentioned earlier is characteristically associated with a rearrangement of the microstructure, whereby interaction effects between elements may reach extreme magnitudes. Thus the behaviour is indicative of yielding in a ductile metallic structure. In a phenomenological theory, this is usually represented in terms of a yield function or yield point on the corresponding stress–strain curve of Fig. 9. For a somewhat longer duration of the transient stage as, for instance, during creep deformations, the microstructure tends to stabilize to the steady-state deformation. In the present theory this type of transient behaviour is therefore seen as a limiting process connecting the end of the reversible with the beginning of the steady-state deformation process. There is, however, another transient stage which relates to the breakdown of the microstructure during a finite time interval. In metallic and composite systems, it is the stage of crack initiation and propagation

4.5 Stability of microstructures

leading to brittle fracture. In fibrous structures of the bonded type this transient involves partial or complete bond breakage resulting ultimately in failure by rupture. Both these classes of discrete materials have been more recently investigated. Thus a somewhat modified form of the probabilistic analysis discussed here has been used in ref. [132] for the modelling of these transients in relation to fracture and fatigue phenomena in metals. Analytical and experimental studies of fibrous systems, in particular with reference to bond failure and the experimental determination of the transition intensities, are discussed in references [86, 91, 92]. The criteria for the limiting process required for the representation of the transient stage of deformations may be stated as follows:

(i) the intensity with which the deformation process $\mathbf{u}(t)$ changes from state i to an adjacent state j in a time interval Δt is:

$$\frac{\lambda t_2(t-t_1)}{t(t_2-t_1)} \Delta t + 0(\Delta t); \quad t_1 \leqslant t \leqslant t_2;$$

(ii) the intensity with which $\mathbf{u}(t)$ changes from a state i to any other state except $j, j = i+1$ in the interval Δt is: $0(\Delta t)$;

(iii) the intensity with which $\mathbf{u}(t)$ does not change from state i to a state j in Δt is:

$$1 - \frac{\lambda t_2(t-t_1)}{t(t_2-t_1)} \Delta t - 0(\Delta t); \quad t_1 \leqslant t \leqslant t_2$$

from which it is seen that the elements of the transition matrix $Q(t)$ will be:

$$\left. \begin{array}{ll} q_{ii}(t) = -\dfrac{\lambda t_2(t-t_1)}{t(t_2-t_1)} & \text{for} \quad i = 0, 1, \ldots, \\[2mm] q_{ij}(t) = \dfrac{\lambda t_2(t-t_1)}{t(t_2-t_1)} & \text{for} \quad j = i+1 \\[2mm] \phantom{q_{ij}(t)} = 0 & \text{otherwise.} \end{array} \right\} \quad (4.68)$$

Introducing to simplify the analysis two new parameters of the type:

$$a = \frac{\lambda t_1 t_2}{t_2 - t_1}; \quad b = \frac{\lambda t_2}{t_2 - t_1} \quad (4.69)$$

gives:

$$\left. \begin{array}{ll} q_{ii}(t) = -b + \dfrac{a}{t} & \text{for} \quad i = 0, 1, \ldots, \\[2mm] q_{ij}(t) = b - \dfrac{a}{t} & \text{for} \quad j = i+1 \\[2mm] \phantom{q_{ij}(t)} = 0 & \text{otherwise.} \end{array} \right\} \quad (4.70)$$

Hence the Kolmogorov differential equation becomes:

$$\left.\begin{array}{c}\dfrac{dP^u_{ij}(t)}{dt} = -bP^u_{ij}(t)+\dfrac{a}{t}P^u_{ij}(t)+bP^u_{i,j-1}(t)-\dfrac{a}{t}P^u_{i,j-1}(t)\\[6pt]\text{for } t_1 \leqslant t \leqslant t_2\end{array}\right\} \quad (4.71)$$

with the solution:

$$\left.\begin{array}{rl}P^u_{ij}(t) = & \dfrac{[b(t-t_1)-a\ln(t/t_1)]^{j-i}}{(j-i)!}\left(\dfrac{t}{t_1}\right)^a e^{-b(t-t_1)}; \quad j \geqslant i\\[6pt]= & 0; \hspace{17em} j < i.\end{array}\right\} \quad (4.72)$$

As in the case of the other stages, $P^u_{ij}(t)$ for the transient stage gives a proper probability measure since $0 \leqslant P^u_{ij}(t) \leqslant 1$ for $\forall i,j$ and $\underline{P}^u(t)$ is a contraction, i.e. $\|\underline{P}^u(t)\| \leqslant 1$. The analysis of the second type of transient behaviour concerning the final stage of the response and the failure of the material is more complex. It is conjectured that the final response stage (fracture) can be modelled by a secondary stochastic process involving second- and higher-order transition probabilities. The latter can be derived from the first-order ones. Such a process would then be a function of the Markov process $\mathbf{u}(t)$, i.e. $\phi\{\mathbf{u}(t)\}$. It is to be noted, that a complete description of fracture in solids necessitates a random field formulation for the solution of a given boundary value problem (see the following discussion in section 4.6). Whilst in general such a process may not satisfy the Markov property or the Kolmogorov relation, there are nevertheless certain conditions that make it possible to satisfy the latter. Since the random deformation process has been defined by $\mathbf{u}_t(\mathbf{X})$ or $\mathbf{u}(t) \in [\mathscr{U}_\infty, \mathscr{F}_\infty, \mathscr{P}^u]$ in which $\mathscr{U}_\infty$ is the ∞-fold product of $\mathscr{U}$ and $\mathscr{F}_\infty$ the ∞-fold product of $\mathscr{F}$, the final stage process $\phi\{\mathbf{u}(t)\}$ will be such that $\mathbf{u}(t) \in \mathscr{U}_\infty \setminus \mathscr{U}_s$. Hence, it will be valid only in the subspace $\mathscr{U}_\infty \setminus \mathscr{U}_s$. If $\mathscr{U}_s$ is the space of $\mathbf{u}(t)$ prior to the final stage in the deformation, then $\mathscr{U}_\infty \setminus \mathscr{U}_s$ will be the space associated with $\phi\{\mathbf{u}(t)\}$ and the breakdown of the microstructure. Since the function $\phi\{\mathbf{u}(t)\}$ can be regarded as a transformation of $\mathbf{u}(t)$, the σ-algebra $\mathscr{F}$ in $\mathscr{U}_\infty \setminus \mathscr{U}_s$ will be generated by ϕ from that in $\mathscr{U}_s$. Further considerations of this stage of deformation are, however, excluded from the present analysis.

The significance of the transition intensities in the representation of the various stages of the general deformation process (eqns. 4.58–4.72) has become apparent. It has been shown with regard to the transient material behaviour that in a general deformation process two types of transient stages may be distinguished. The first type shows the tendency of $q_{ij}(t) \to q_{ij}|_s$ as the variable time $t \to t_s$, the latter being the instant of time when the steady-state deformations are first observed. In accordance with form (4.50) and on the assumption that $P_{ij}(t)$ is t-continuous on $X(\mathscr{U})$ for $\forall t, s \in \mathbb{R}^+$ a limit will be reached such that:

$$\lim_{t \to s}[\underline{P}(t)-\underline{P}(s)]\mathscr{P}^u(0) \to 0 \quad \text{for all } s > 0. \quad (4.73)$$

4.5 Stability of microstructures

This characteristic will prevail during the stable steady-state deformation stage until a time $t < t_f$ is reached. The time t_f corresponds then to the beginning of the second type transient, i.e. the onset of the microstructural breakdown. The stable stage preceded by the first transient can also be reached analytically, by considering a Cauchy sequence $\{P(t_n), n = 1, 2, \ldots\}$ in $L(0, 1)$ for which the following convergence property must be required:

$$|P_{ij}(t_{n+1}) - P_{ij}(t_n)| < \varepsilon \text{ for } n > N(\varepsilon) \qquad (4.74)$$

in which $0 < \varepsilon t \ll 1$ corresponds to a small time interval. By comparison with eqn. (4.50), it is seen that the above condition expresses the same requirement for a stable deformation to occur, i.e. that $0 < q_{ij} \ll 1$. The distinct features of stability of the microstructure of a discrete material in terms of the characteristic transition intensities are summarized in Table 1 (see also ref. [136]).

TABLE 1. Regions of microstructural stability
(λ_{t_r} = transient, λ_s = steady-state, λ_f = final stage intensity)

Deformation stage $\mathbf{u}(t)$	Elements $q_{ij}(t)\ \mathbf{Q}(t)$	State i, j	Time interv. Δt	Trans. intensity λ	Stability of m. struct.
Elastic deformations	$q_{ii} = 1$ $q_{ij} = 0$	$i = 0, 1 \ldots$ $j = i + 1$	$0 \to t_1$	$\lambda = 0$	stable
Transient deformations	$q_{ii}(t) =$ $-b + a/t$	$i = 0, 1, \ldots$	$t \to t_1^+$	$0 \to \lambda_{t_1}^+$	unstable
	$q_{ij}(t) = b - a/t$	$j = i + 1$	$t_1 \to t_2$	$\lambda_{t_r} > \lambda_s$	unstable
	$q_{ij}(t) = 0$	otherwise		$\lambda_{t_r} \equiv \lambda_s$	stable
Steady-state deformations	$q_{ii} = -\lambda_s$	$i = 0, 1 \ldots$ $j = i + 1$	$t_2 < t_f$	$0 < \lambda_s \ll 1$	stable
	$q_{ij} = \lambda_s$		at t_f	$\lambda_s \to \lambda_f$	unstable
Final trans. (breakdown of m. struct.)		$j = i + 1,$ $i + 2,$ $\vdots$	t_f	$\lambda_s > \lambda_f$ $\lambda_f \to \infty$	unstable

From the above presented general deformation theory of discrete solids and the considerations given to the microstructural stability the following statements can be made:

(i) A random deformation process of structured solids in a probability space [$\mathcal{U}$, $\mathscr{F}^u$, $\mathscr{P}^u$] is a sequence of the random variables $\{u(X, t)\}$ or $\{u(t)\}$ each member of this sequence being generated from the previous one by a random endomorphism.

(ii) The general probabilistic deformation process can be subdivided into two stable stages that correspond to the purely elastic and steady-state response of the material. The intermediate and final stage in the deformation process are transient, whereby the former tends to approach the conditions for stability of the microstructure and the latter is unstable leading to failure of the material.

(iii) The most important characteristic in the representation of the process is the transition probability and the associated transition intensity. The transition probabilities may be regarded as the kernels of evolution operators on the deformation space as a subspace of $\mathscr{X}$.

(iv) For stability to exist the transition probabilities must satisfy the semigroup property. This property is not satisfied for the two transient stages mentioned above.

It is of interest to note that the above division of stability zones in a structured solid finds an immediate application in the more recent analysis of crack propagations in metals (see, for instance, Hollstein [137]).

4.6 MATERIAL OPERATORS AND BOUNDARY-VALUE PROBLEMS

Material operators according to Postulate 3 of the probabilistic mechanics theory have been discussed in subsection 3.6.2 of Chapter 3. Such operators reflect the random physical properties of a solid on the one hand and on the other provide the connection between the stress and deformation subspaces of the general state space. Since the mapping between these spaces is intimately related to a given boundary value problem, the significance of these operators as random operators should be further emphasized. Thus either of these operators, i.e. the micro-, meso- and macro-material operator, lead to random operator equations that will be discussed subsequently. In order to show the application of such operators in general to the classical forms of boundary value problems in the theory of elasticity, i.e. the Dirichlet and Neumann problems, the latter can be stated as follows:

(i) *the Dirichlet problem*:

given a linear operator L on a Hilbert space $\mathcal{U}$, a specified function (body force) $f \in F$ and a prescribed deformation on the boundary $g \in \partial \mathcal{U}$, the problem consists of finding the deformation $u \in \mathcal{U}_L$ such that:

$$\left. \begin{array}{l} L u = f \quad \text{in } \mathcal{M} \text{ (macrodomain)}, \\ u = g \quad \text{on } \partial \mathcal{M}. \end{array} \right\} \quad (4.75)$$

4.6 Material operators and boundary-value problems

(ii) *the Neumann problem:*

given a linear operator L *on the deformation space* $\mathscr{U}$ (Hilbert space), a specified function (body force) $\mathbf{f} \in F$ and a prescribed normal derivative of $\mathbf{u}$ on $\partial \mathscr{M}$ the problem consists of finding the deformation $\mathbf{u} \in \mathscr{U}_L$ such that:

$$\left. \begin{array}{l} L\mathbf{u} = \mathbf{f} \quad \text{in} \quad \mathscr{M}, \\ \dfrac{\partial \mathbf{u}}{\partial \mathbf{n}} = \mathbf{s} \quad \text{on} \quad \partial \mathscr{M}, \end{array} \right\} \quad (4.76)$$

Evidently, the above operator L in both these problems in the deterministic formulation will have its analogous form in the probabilistic theory. It is to be noted, however, that the primary function of the operator L in continuum mechanics is to connect the field equations (stress equations of motion) with the constitutive relations and thus to reduce the problem to one for the deformations only. This operation in the probabilistic mechanics theory is not so simple. Although the stress-deformation and or stress–strain relations are linked by the material operator, the relevant field equations may not always be available. Hence it becomes necessary in the probabilistic analysis to resort to an operational formalism, i.e. to define boundary-value problems in terms of random operators. This leads then to random operator equations, the solutions of which cannot be obtained generally in a closed form so that approximate methods of solutions are required. A comprehensive discussion on this subject matter is given by Bharucha-Reid [74], Kannan [138], Chow [139] amongst others. To obtain the probabilistic forms analogous to the deterministic ones, for the above stated boundary-value problems, a structured solid may be considered for the simplicity of the analysis as a random medium. In particular the following assumptions are made:

(i) the discrete medium is replaced by a continuous random medium;
(ii) the physical domain for an ensemble of microelements corresponds to an equivalent domain in the random continuum medium such that the statistical properties are preserved and no discontinuities are taken into account;
(iii) the field quantities are considered in the classical sense.

As a consequence of these assumptions, the field quantities are considered to act at mathematical points identified by the corresponding position vectors X_i in the medium. Thus the conventional stress equation of motion for the elasto-dynamic case is given by:

$$\sigma_{ij,j} + \rho(\omega) f_i = \frac{\partial^2 u_i}{\partial t^2} \quad (4.77)$$

and in the elasto-static case by:

$$\sigma_{ij,j} + \rho(\omega)f_i = 0 \tag{4.78}$$

where f_i designates the body force vector and σ_{ij} the Cauchy stress tensor. The mass density $\rho(\omega)$ is considered as a random variable. If the material operator (macroscopic) is used, in the sense of continuum mechanics which connects the stress and deformation subspaces Σ, $\mathcal{U} \subset \mathcal{X}$, then:

$$M(\omega): \Sigma \to \mathcal{U}, M(\omega) \in \mathcal{L}(\Sigma, \mathcal{U}) \tag{4.79}$$

and hence
$$M(\omega)\sigma(\omega) = \mathbf{u}(\omega) \tag{4.80}$$

where $M(\omega)$ is a bounded linear operator for a purely elastic medium. In this case $M^{-1}(\omega)$ will exist and can be given explicitly by:

$$M^{-1}(\omega) = E(\omega) \cdot \text{grad} = E_{ijkl}\frac{\partial}{\partial x_l} \tag{4.81}$$

in which E is the modulus of elasticity. Employing this form (4.81) in (4.78) gives the deformation equations of motion in operational form as:

$$L(\omega)\mathbf{u} + \rho(\omega)f = 0 \tag{4.82}$$

where
$$L(\omega) = \text{div}(\mathbf{E} \cdot \text{grad}) = \frac{\partial}{\partial x_i}\left(E_{ijkl}\frac{\partial}{\partial x_l}\right). \tag{4.83}$$

To formulate the boundary-value problems (i) and (ii) it is necessary to introduce the following function spaces:

(i) $\mathcal{U}_L = \{\mathbf{u} \in \mathcal{U}; L(\omega)\mathbf{u} \in F\}$;
(ii) $\partial \mathcal{U}$ = space of boundary values of $\mathbf{u}$ at the boundary of the specified domain $\mathcal{M}$, i.e. $\partial \mathcal{M}$;
(iii) ∂G = space of boundary values for $\partial \mathbf{u}/\partial \mathbf{n}$ at $\partial \mathcal{M}$, $\mathbf{n}$ being the outward normal to the boundary surface;
(iv) ∂G^* = the dual space of ∂G.

By using these definitions both the Dirichlet and Neumann boundary-value problems of the theory of elasticity can be formulated in their probabilistic forms in terms of the following operator equations:

(i) *The Dirichlet problem for $L(\omega)$:*
given $\mathbf{f} \in F$ and $\mathbf{g} \in \partial \mathcal{U}$ find $\mathbf{u} \in \mathcal{U}_L$ such that

$$\left. \begin{array}{c} L(\omega)[\mathbf{u}] = \mathbf{f} \\ \mathbf{u}|_{\partial \mathcal{M}} = \mathbf{g} \end{array} \right\} \tag{4.84}$$

4.6 Material operators and boundary-value problems

(ii) *The Neumann problem for* $L(\omega)$:
given $\mathbf{f} \in F$ and $\mathbf{s} \in \partial G^*$, find $\mathbf{u} \in \mathscr{U}_L$ such that:

$$\left. \begin{array}{r} L(\omega)[\mathbf{u}] = \mathbf{f}, \\ \dfrac{\partial \mathbf{u}}{\partial \mathbf{n}} \bigg|_{\partial \mathscr{M}} = \mathbf{s}. \end{array} \right\} \quad (4.85)$$

It is seen that relations (4.84, 4.85) represent the analogous forms of the classical ones in elasticity theory. However, if the solid is regarded as a structured medium, this formulation has to be extended by a generalization of the random operators and the corresponding equations. For this purpose recalling the definition of Banach space-valued random variables given in section 1.3 (ii) of Chapter 1, the random operators discussed above can be expressed in terms of such variables. Such an approach for defining random operators and transformations in Banach spaces has been given by Bharucha-Reid [140], Browder [141], Kannan [138], Adomian [142] amongst others. In accordance with the principles of probabilistic mechanics these operators can be defined in the following manner:

Def. 1: If x is a random variable with values in a separable Banach space X and T an a.s. continuous operator of $\mathscr{X} \times X$ into another Banach space Y, the mapping $T(\omega)$ is a random operator, $\omega \in \mathscr{X}$.

Thus considering the stress and deformation subspaces Σ, $\mathscr{U} \subset \mathscr{X}$ such a random operator will connect these spaces by:

$$T(\omega)\,\sigma = \mathbf{u}(\omega) \quad \text{for every} \quad \sigma \in \Sigma. \quad (4.86)$$

Hence, it is seen that this random operator is a material operator M given by Postulate 4 of the probabilistic theory. However, in the fundamental definitions, the material operator has been given as a micro-, meso- and macro-operator according to its belonging to a micro-element, a meso-domain or a macro-domain of the material body. It follows that not only the constitutive relations, but also the boundary-value problems will have to be specified on the boundary of these domains.

Since in the deterministic theory boundary values in general refer to the macroscopic material domain analogously in the probabilistic formulation one can use the macro-material operator. However, it is fundamental in the present theory to employ the random micro-operator that requires an analysis in terms of a random field formulation. The same holds for the application of the meso-material operator, but with a random field of a more approximate structure. In this context, it is to be noted that the random theory (see, for instance, [143]) is still in an active stage of development and hence its application to the solution of boundary-value problems cannot be given as yet. However, it is possible to

114 *Probabilistic Mechanics of Solids*

consider evolutions of the above discussed random operators [74]. To this extent random transformations corresponding to such an evolution as shown previously for the deformation space (section 4.4) can be used by defining a random endomorphism $T(\omega): \mathscr{X} \times \mathscr{U} \to \mathscr{U}$.

4.7 DYNAMICS OF STRUCTURED SOLIDS

In this section the wave propagation through solids with a random structure will be considered. Earlier work by Kampé de Feriét [144], Keller [145], Frisch [146] and [147–149] dealt with the wave mechanics of random continuous media. This work was concerned with the formulation of the wave propagation in random media and the solution of the corresponding random differential and integral equations. The aspect of discreteness of the microstructure other than that of a perfect lattice, however, has not been considered. Hence the present analysis will be carried out on the basis of the principles of probabilistic mechanics. In the first part of this section definitions are given with reference to the involved field quantities of the 1-D wave propagation in a semi-infinite bar. It is important to note, that for this purpose the spatial scales according to Postulate 1 of the theory are extended to allow also time scales in the formulation of the wave phenomena. Thus, a macroscopic time for the passage of the wave through the entire bar and a microtime with reference to the wave propagation through an individual element of the microstructure will be introduced. It is apparent that due to the discreteness of the microstructure the wave motion through the macroscopic material body will occur in such a manner that any wave front passing through the individual elements will bring about changes in its direction, amplitude and velocity of propagation. The complexity arising from changes in direction in the general case are considerably reduced by dealing first with the one-dimensional problem [150].

(i) *Longitudinal wave propagation (1-D model)*:

In order to simplify the analysis the physical domain of the macroscopic body is seen as a sequence of cubic microelements. For the formulation of the wave propagation the intrinsic random field quantity is taken as the wave velocity vector ${}^\alpha \mathbf{v} \in \mathscr{V} \subset \mathscr{X}$, which is also the deformation rate of a given microelement. Since from the onset the interfaces between contiguous elements are admitted, although they may be small, the definition of the following field parameters is required:

Def. 1: The reflection coefficient: $\quad C_r \stackrel{df}{=} \dfrac{{}^\alpha \mathbf{v}_r}{{}^\alpha \mathbf{v}_i},$

$\qquad$ the transmission coefficent: $\quad C_{tr} \stackrel{df}{=} \dfrac{{}^\beta \mathbf{v}_{tr}}{{}^\alpha \mathbf{v}_i},$

$\hfill (4.87)$

4.7 Dynamics of structured solids

where $^{\alpha}\mathbf{v}_i$, $^{\alpha}\mathbf{v}_r$ and $^{\beta}\mathbf{v}_{tr}$ are the incident, reflected and transmitted wave velocity vectors, respectively.

Def. 2: The passage time $^{\alpha}\tau \stackrel{df}{=} {}^{\alpha}d/{}^{\alpha}c$ is the time of passing of a wavefront through an element α, where $^{\alpha}d$ is the size of the element in the direction of propagation and $^{\alpha}c$ the wave-propagation velocity.

The above definitions make it possible to account for two important effects arising in the wave propagation through the structured medium, i.e.:

(i) the transmitted wave velocity $^{\beta}\mathbf{v}_{tr}$ is different from the incident wave velocity $^{\alpha}\mathbf{v}_i$ due to the existence of the ($\alpha\beta$) grain boundary;
(ii) the wave propagation velocity $^{\alpha}c$ is random in every micro-element.

It is seen that Definition 2 introduces a time-wise scale effect due to the admittance of a microstructure, which is dependent on the size of the micro-element itself.

Thus individual wave-propagation velocities can be assumed to occur in a structured solid. It is to be noted that the distributions of the transmission coefficient C_{tr} and passage time $^{\alpha}\tau$ are accessible from experimental investigations on single and or bi-crystals, respectively. However, if such data are not available these distributions can be obtained by a simulation technique such as the Monte Carlo method [151]. It is apparent from the above given definitions that the wave velocity field is a random field. This necessitates to regard the time-scales belonging to $\mathscr{I}$ of the Markov random field represented by $\mathscr{V}$ ($\mathbb{R}^1 \times \mathscr{I}$). The two dimensionality of $\mathscr{I} = \mathscr{I}_1 \times \mathscr{I}_2$ is readily recognized from the following definitions:

Def. 3: $\bar{t}$ is the macroscopic (average) time for the average path of a longitudinal wave, $\bar{t} \in \mathscr{I}_1$, t is the internal time for a sample path of a longitudinal wave, $t \in \mathscr{I}_2$.

Due to the random velocity $^{\alpha}c_L$ the longitudinal wave arrives anywhere in a time-space graph with an indeterminate time difference with respect to the macro-disturbance (average path). Following the above definitions a special random variable can be identified:

Def. 4: $\tau_L \stackrel{df}{=} t - \bar{t}$ is the dispersion time for the longitudinal wave.

Whilst τ_L is a random variable intrinsic in the $\mathbf{v}(\bar{t}, t)$ process, it itself is a random process parametrized by the average time $\bar{t}$ so that:

$$\tau_L = \tau_L(\bar{t}). \tag{4.88}$$

Thus, one can write that:

$$P\{\mathbf{v}(\bar{t}, t)\} = P\{\mathbf{v}(t)\} P\{\tau_L(\bar{t})\}. \tag{4.89}$$

It is evident from the above relation that both effects (i) and (ii) have been separated. Thus the first term on the right-hand side of (4.89) represents a modulation of the wave-velocity vector, whilst the second term accounts for the dispersion of the wavefront in a time-space graph. Moreover, this relation (4.89) indicates that the velocity v evolves independently when parametrized by the internal and the average times. These evolutions will now be recognized as Markov processes. Considering first the L-wave evolution in terms of the internal time t and writing:

$$\mathscr{P}\{\mathbf{v}(t)\in E\} = \int_{\mathscr{V}} \mathscr{P}\{\mathbf{v}(t)\in E\,|\,\mathbf{v}(t_0) = x\}\,\mathscr{P}(\mathrm{d}x),\ t_0 \leqslant t \qquad (4.90)$$

it expresses the Markov property of the $\mathbf{v}(t)$ process. Writing the conditional probability as an explicit function of the probability distributions of the transmission coefficient and the passage time then:

$$P\{\mathbf{v}(t_0+{}^\alpha\tau)\in E\,|\,\mathbf{v}(t_0) = x\} = \mathscr{P}\{C_{t_r}:\mathbf{v}(t_0+{}^\alpha\tau) = C_{t_r}\cdot\mathbf{v}(t_0)\in E\} \qquad (4.91)$$

Recognizing this Markov process to be a temporally homogeneous one and since $\mathscr{V}$ is a subspace of $\mathscr{X}$ and is separable, one can postulate the existence of a transition function given by:

$$P({}^\alpha\tau,x,E)\stackrel{\mathrm{df}}{=}\mathscr{P}\{\mathbf{v}({}^\alpha\tau)\in E\,|\,\mathbf{v}(0) = x\},\quad {}^\alpha\tau\in\mathscr{I}_2 = [0,\infty) \qquad (4.92)$$

satisfying the usual conditions (see Chapter 2).

Hence the corresponding Chapman–Kolmogorov equation will be given by:

$$\left.\begin{array}{c} P(t+t',x,E) = \displaystyle\int_{\mathscr{V}} P(t,x,\mathrm{d}y)\,P(t',y,E);\quad t = \displaystyle\sum_{\alpha=1}^{n} {}^\alpha\tau, \\[2mm] t' = \displaystyle\sum_{\alpha=1}^{n'} {}^\alpha\tau \geqslant 0 \end{array}\right\} \qquad (4.93)$$

Let $C(\mathscr{V})$ be a Banach space of all bounded continuous functions $f(x)$ on the velocity space $\mathscr{V}$ endowed with the norm $\|f\| = \sup_{x\in\mathscr{V}}|f(x)|$. One can then define an operator ${}^2T$ on $\mathscr{V}$ as follows:

$${}^2T(t)[f(x)]\stackrel{\mathrm{df}}{=}\int_{\mathscr{V}} f(y)\,P(t,x,\mathrm{d}y);\quad f\in C(\mathscr{V}). \qquad (4.94)$$

It follows from section 2.2 (ii) in Chapter 2 that $\{{}^2T(t), t \geqslant 0\}$ is a contraction semi-group of operators on $C(\mathscr{V})$.

Returning to eqn. (4.89) and considering the second term on the right-hand side of it that involves the L-wave propagation, the corresponding stochastic process $\mathbf{v}(\bar{t},t) = \mathbf{v}(\bar{t})$ for a fixed time t will be characterized by the following probability function:

$$P\{\mathbf{v}(\bar{t},t)\} = \mathrm{const}\cdot P\{\tau_L(\bar{t})\}. \qquad (4.95)$$

4.7 Dynamics of structured solids   117

Consequently the evolution of this process corresponds to a linear birth–death Markov process at regular intervals in terms of the average passage time $\langle {}^\alpha \tau \rangle$. This process can be represented by a one-step transition function given by:

$$P(\bar{t}_0, z, \bar{t}, D^L) \stackrel{df}{=} \mathscr{P}\{\tau_L(\bar{t}) \in D^L \,|\, \tau_L(\bar{t}_0) = z\}, \quad \bar{t} \in \mathscr{I}_1 \qquad (4.96)$$

which is a function of the distribution of the passage time $P({}^\alpha \tau)$ so that:

$$\mathscr{P}\{\tau_L(\bar{t}) \in {}^L D \,|\, \tau_L(\bar{t}_0) = z\} = P\{{}^\alpha \tau : {}^\alpha \tau - \langle {}^\alpha \tau \rangle = \tau_L(\bar{t}) - \tau_L(\bar{t}_0)\}, \qquad (4.97)$$

where $\bar{t} = \bar{t}_0 + \langle {}^\alpha \tau \rangle$,

τ_L is a random variable in the state space $\mathscr{D}^{\tau_L}$ of the dispersion times, and D^L is a Borel set in $\mathscr{D}^L$ belonging to the Borel field $\mathscr{F}^L$.

Heuristically this process can be seen as a temporally homogeneous one, so that the following transition function can be adopted:

$$P(\bar{t}, z, D^L) \equiv P(0, z, \bar{t}, D^L) \qquad (4.98)$$

which satisfies the standard conditions given in Chapter 2. Hence the corresponding Chapman–Kolmogorov equation assumes the following form:

$$P(\bar{t}+\bar{t}', z, D^L) = \int_{\mathscr{D}^L} P(\bar{t}, z, dw) \cdot P(\bar{t}', w, D^L); \quad \bar{t}, \bar{t}' \geq 0. \qquad (4.99)$$

Since from (4.97) the $\tau_L(\bar{t})$ process is time-homogeneous a transition operator ${}^1T$ parametrized by the average time $\bar{t}$ can be defined in the following form:

$${}^1T(\bar{t})[f(z)] \stackrel{df}{=} \int_{\mathscr{D}^L} f(w) P(\bar{t}, z, dw); \quad f \in C(\mathscr{D}^L). \qquad (4.100)$$

It follows again from section 2.2 (ii) of Chapter 2 that $\{{}^1T(\bar{t}), \bar{t} \geq 0\}$ is a contraction semi-group of the operators ${}^1T$ on $C(\mathscr{D}^L)$.

It is now necessary to combine the above results in obtaining the two transition operators for the description of the L-wave propagation in a discrete 1-D medium, by introducing the following theorem:

Theorem 1 (M. Ostoja-Starzewski [152]):

"If $\{{}^2T(t), t \in \mathscr{I}_2\}$ and $\{{}^1T(\bar{t}), \bar{t} \in \mathscr{I}_1\}$ are one-parameter semi-groups of contraction operators on $C(\mathscr{X})$, then there exists a two-parameter semi-group $\{T(\bar{t}, t), (\bar{t}, t) \in \mathscr{I}_1 \times \mathscr{I}_2\}$ such that:

(i) $T(\bar{t}, t)$ is the direct product of two operators ${}^1T$ and ${}^2T$ so that $T(\bar{t}, t) = {}^1T(\bar{t}){}^2T(t)$;

(ii) $\lim\limits_{(\bar{t},t) \to 0} T(\bar{t}, t) = I$;

(iii) $T(\bar{t}, t)$ is a contraction: $\|T(\bar{t}, t)\| \leq 1$."

From the above brief discussion, it can be seen that the evolution of the longitudinal wave propagation in a one-dimensional model can be described by an abstract dynamical system or the quadruplet $[\mathscr{V} \times \mathscr{D}^L, \mathscr{F}^v \times \mathscr{F}^L, \mathscr{P}^v \times \mathscr{P}^L, T(\bar{t}, t)]$. The semi-group property indicates that both Chapman–Kolmogorov relations (4.93) and (4.99) can be jointly used to obtain the evolution of the velocities $\mathbf{v}(\bar{t}, t)$. In this context it is to be noted, that the theorem concerning multi-parameter semi-groups is due to Hille and Phillips [118] in the classical functional analysis. The extension for the application to Markov fields and the proof is given in [152].

(ii) *Uni-directional wave-propagation in solids* (*3-D model*)

To formulate the problem of random plane waves in a three-dimensional structured solid the same simplification is made as before, i.e. by considering only a cubic structure with random physical properties. It is assumed that initially at the end face of the body (semi-infinite bar) plane waves are generated by the application of a uniform pulse. Due to the random physical properties of the elements of the structure this plane wave will evolve into a random plane wave. Although the interactions between contiguous elements occur in all three directions the dominant interaction will be in the direction of propagation of the wave. It is assumed that the macroscopic domain of the bar can be regarded to consist of a number of sequences each of which contains a denumerable number of microelements. In this sense the wave propagation consists predominantly of a longitudinal wave with slight disturbances in the passage through individual microelements. It is further considered that the coupling between the sequences is weak with respect to the wave propagation. Similarly to the case of the one-dimensional model, two effects will be significant here, i.e.:

(i) the transmitted wave velocity $^\beta v_{tr}$ is different from the incident wave velocity $^\alpha v_i$ due to the grain boundary $\partial^\alpha \mathscr{D}$ ($^\alpha \mathscr{D}$-domain of a single grain).
(ii) the wave propagation velocity $^\alpha c$ is random in every microelement α.

It is to be noted that the velocity $^\alpha \mathbf{v}$ goes into $^\beta \mathbf{v}$ after the random passage time, which in this case has to account for the transmission process corresponding to the interaction with four neighbours (cubic structure). Hence the wave velocity vector $\mathbf{v}$ becomes a function of the three Euclidean coordinates. It is obvious that here a transmission operator will be required rather than a transmission coefficient. The former is defined as follows:

Def. 5: The transmission operator is a random mapping:

$$C(\omega): {}^\alpha \mathbf{v}(t) \to {}^\beta \mathbf{v}(t + {}^\alpha \tau)$$

where $^\alpha \mathbf{v}, {}^\beta \mathbf{v} \in \mathscr{V}$.

4.7 Dynamics of structured solids

Generally, this operator depends on the surface interactions at $\partial^\alpha \mathscr{D}$ and the velocities in the four neighbouring grains $\gamma_i, i = 1, \ldots, 4$, i.e.:

$$C = C(C_{\text{tr}} \text{ on } \partial^\alpha \mathscr{D}, {}^{\gamma_i}\mathbf{v}; i = 1, \ldots, 4). \tag{4.101}$$

It is to be noted that the ${}^{\gamma_i}\mathbf{v}$'s are time-dependent random processes, which makes the above defined random operator also time-dependent. It has been shown in ref. [152] that the transmission operator $C(\omega)$ is a random endomorphism on $\mathscr{V}$ and for the elastic range is an $\mathscr{L}(\mathscr{V})$-valued random function. It has also been shown that both effects (i) and (ii) can still be formulated by Markov processes, however their joint effect in the wave propagation has to be established in terms of a Banach space-valued random field. To this effect the following definition is given:

Def. 6: A Banach space-valued random field on $\mathscr{I} = \mathscr{I}_1 \times \mathscr{I}_2 \times \{{}^\alpha\mathscr{D}\}$ and ${}^\alpha\mathscr{D} \subset \mathbb{R}^3$ is a mapping $\mathbf{v}\,(\bar{t}, t, {}^\alpha\mathbf{X}, \omega) \equiv \mathbf{v}\,(s, {}^\alpha\mathbf{X}, \omega)$: $\mathscr{I} \times \mathbb{R}^3 \times \Omega \to \mathscr{V}$ such that for every $s \in \mathscr{I}$ and $\mathbf{X} \in {}^\alpha\mathscr{D}$, $\mathbf{v}$ is a Banach space-valued random variable; $\mathscr{V}$ is the velocity subspace of $\mathscr{X}$.

Although the above representation is rigorous, it is convenient to use a simplification by employing a real valued random field instead and introducing the notion of a power flux. The latter can be expressed on the assumption that each microelement is considered as a continuum. Thus it is known from continuum theory that the power flux ${}^{\gamma_i}\pi$ (energy identity) is given by:

$$ {}^{\alpha\gamma_i}\pi = \frac{\Delta^{\alpha\gamma_i}K}{{}^\alpha\tau} + \frac{\Delta^{\alpha\gamma_i}U}{{}^\alpha\tau} = \frac{\Delta^{\alpha\gamma_i}K}{{}^\alpha\tau} + \frac{\Delta^{\alpha\gamma_i}U\,{}^\alpha c}{{}^\alpha d}; \quad i = 1 \ldots 4 \tag{4.102}$$

where K is the kinetic energy and U the elastic potential of the grain participating in the wave motion. If the flux is taken as a linear functional, i.e. ${}^{\alpha\gamma_i}\pi = f({}^\alpha\mathbf{v}, {}^{\gamma_i}\mathbf{v})$, then the total change of energy in the microelement α due to the interaction with four neighbours becomes:

$$ {}^\alpha\pi = \sum_{i=1}^{4} {}^{\alpha\gamma_i}\pi = \sum_{i=1}^{4} f({}^\alpha\mathbf{v}, {}^{\gamma_i}\mathbf{v}) = \tilde{f}({}^\alpha\mathbf{v}) \tag{4.103}$$

In this relation ${}^\alpha\pi$ in a given sequence S_k is a function of the $\mathbf{v}\,(\bar{t}, t)$ process. This means that it is a generalized random functional ${}^\alpha\pi: {}^\alpha\mathscr{V} \times \mathscr{D}^L \to \mathbb{R}^+$ in the sense indicated previously in section 4·4 of this chapter. Thus by definition 6, the entire wavefront is described by a real-valued generalized random field (meso-domain) $\pi: {}^{\mathscr{M}}\mathscr{V} \times \mathscr{D}^L \to \mathbb{R}^+$.

It is evident from Theorem 1 that the evolution of the disturbances in a given sequence can be expressed by:

$$T(\bar{t},t)[^{\alpha}\mathcal{P}(z,x)] = {}^1T(\bar{t})\,{}^2T(t)[^{\alpha}\pi(z,x)] =$$

$$= \int_{\mathscr{V}}\int_{\mathscr{D}^L} {}^{\alpha}\pi(w,y)P(\bar{t},z,dw)P(t,x,dy), \qquad (4.104)$$

in which the functional on $C(\mathscr{D}^L \times {}^{\alpha}\mathscr{V})$ will be given by:

$$^{\alpha}\pi(z,x) \equiv {}^{\alpha}\pi(\tau_L(\bar{t}_0),\mathbf{v}(t_0))$$

Since in the present analysis the wavefront field $\mathbf{v}(\bar{t},t)$ is assumed to be Markovian, it can be shown that for the transmission from one microelement to the adjacent one, in a given sequence of the cubic structure, the following relation can be written:

$$\left.\begin{array}{r} T(\langle{}^{\alpha}\tau\rangle,{}^{\alpha}\tau)[^{\alpha}\pi(z,x)] \\[4pt] = E\{{}^{\alpha}\pi(\tau_L,\mathbf{v})\text{ at }(\langle{}^{\alpha}\tau\rangle,{}^{\alpha}\tau)|\ \mathscr{F}^L(0,\bar{t}_0)\times\mathscr{F}^{\mathbf{v}}(0,t_0)\} \end{array}\right\} \qquad (4.105)$$

where $\mathscr{F}^L(0,\bar{t}_0) = \prod_{\alpha=1}^{n} \mathscr{F}^L(\langle{}^{\alpha}\tau\rangle)$ and $\mathscr{F}^{\mathbf{v}}(0,t_0) = \prod_{\alpha=1}^{n} \mathscr{F}^{\mathbf{v}}({}^{\alpha}\tau)$ with the initial condition:

$$^{\alpha}\pi(z,x) \quad \text{at times} \quad (\bar{t}=\bar{t}_0, t=t_0).$$

It is readily recognized from the above relation, that depending on the interaction effects due to the immediate neighbours of the single grain α the expected value of the power flux may be larger or smaller than that stated above for a given initial condition. However, it is more important to arrive at a formulation which represents the power flux of the entire wavefront from all participating sequences. In this case it is apparent that due to some energy loss the finally evolved power flux in the system will be smaller or equal to that given by the initial condition. Thus for the evolution of the entire wavefront in the macroscopic time $t=\bar{t}$, it can be formally expressed by:

$$T(\bar{t},\bar{t}) = I^2 T(\bar{t})$$

so that the operational equation (4.105) becomes:

$$^2T(\bar{t})[\pi(x)] = E\{\pi(\mathbf{v}(\bar{t}))|\mathscr{F}^{\mathbf{v}}(0,\bar{t}_0)\} \leqslant \pi(x) \text{ at } \bar{t}_0 \qquad (4.106)$$

This relation shows the super-harmonicity of π under the operator ^2T. This result indicates on the basis of Syski's lemma 5.3 [153], that the total power flux in the Markovian modelling of the wave propagation is a non-negative supermartingale of the Markov process [152]. This also connects the present formulation with the abstract potential theory of Markov processes [154, 46].

5
Probabilistic Mechanics of Fluids

5.1 INTRODUCTION

It is well known that externally applied disturbances to a fluid system are damped due to internal relaxation processes within its microstructure. In a macroscopic description the occurring phenomena are diffusion, viscous flow and thermal conduction processes, which determine the transport characteristics of the fluid. Even without external influences there are, at a given temperature, microscopic fluctuations that are dissipated similarly to the external perturbations. In a fluid, thermal fluctuations occur in a natural manner with a distribution of wavelength and frequencies. For long-wavelength and low-frequency fluctuations, the fluid behaves like a continuum and hence the response can be formulated in terms of the hydrodynamic relations. However, at wavelengths comparable to the intermolecular distance, the molecular structure of the fluid becomes significant and the description of the flow must occur in terms of ensembles of interacting molecules. This approach is usually referred to as molecular fluid dynamics.

The equations of conventional hydrodynamics in general are valid for disturbances of an arbitrary magnitude, if in the corresponding relations terms of momentum and energy density gradients of higher than second order can be neglected. This means that the macroscopic formulation is restricted to a space–time variation, which is slow compared with the scale of molecular processes. Thus the domain of hydrodynamics is that of low wave numbers k and frequencies ω ($2\pi/k$) being the wavelengths of variations.

In the molecular dynamics description several different formalisms have been used (see, for instance, Hansen and McDonald [72], McDonald [155], Boon and Yip [156]). One can distinguish two groups of formulations. In the first group the space–time correlation functions are established by deriving an approximate equation in position and time variables. A typical example of this approach is the use of the generalized Langevin equation that may be regarded as an extension of the linearized equations of hydrodynamics. The second group deals with the same space–time correlation functions and the derivation of an approximate transport equation in position, momentum and time

122 Probabilistic Mechanics of Fluids

variables. This approach uses in general, a linearized Boltzmann equation and is referred to as kinetic theory of fluids. Both these formalisms are essentially the same except for the type of approximation used in the analysis.

In probabilistic mechanics of fluids a different approach will be discussed subsequently in which the geometric and physical characteristics of the fluid medium are introduced as random variables from the onset. Since the probabilistic formulation also requires the description of the evolution of physical and configurational variables with time and the associated probability measures, the notion of an abstract dynamical system introduced in the analysis of solids can also be conveniently used here. This again needs a formalism based on topological and measure theoretical concepts. In the following the fundamental postulates and definitions given in Chapter 3 will be used again, but with the special interpretation pertaining to fluids. Before discussing these concepts and definitions, it may be instructive to review briefly some of the basic relations of statistical mechanics of fluid media.

5.2 MOLECULAR DYNAMICS OF SIMPLE FLUIDS

Due to the limited scope of the present study, the subsequent analysis will be restricted to fluids that are usually termed simple liquids. These are fluids, which are primarily composed of spherical molecules that are chemically inert. Typical for this class of fluids are liquefied rare gases and alkali metals. The behaviour of such media can be well modelled by assuming that the occurring interactions are due to interatomic potentials, which are spherically symmetric and pairwise additive.

(i) *Statistical mechanics consideration* (*time and ensemble averages*)

From a thermodynamics point of view, the properties of a molecular system can be expressed in terms of averages of certain functions of the coordinates and momenta of the molecules contained in that system. In particular for the thermodynamic equilibrium these averages are independent of time. Thus, assuming for simplicity that all molecules are identical and have three degrees of freedom each, as for instance, in a one-component monatomic fluid, the total energy in the absence of an external field can be expressed in terms of the Hamiltonian as follows:

$$\mathcal{H} = \sum_{\alpha=1}^{N} \frac{1}{2\mu} |^{\alpha}\mathbf{p}(t)|^2 + U_N[^{\alpha}\mathbf{r}(t), \ldots, {}^{N}\mathbf{r}(t)] \tag{5.1}$$

in which $^{\alpha}\mathbf{r}, \ldots, {}^{N}\mathbf{r}, {}^{\alpha}\mathbf{p}(t), \ldots, {}^{N}\mathbf{p}(t)$ are the position and momentum vectors of the molecules $\alpha = 1, \ldots, N$, $^{\alpha}\mu \equiv \mu$ respectively and U_N the potential energy due to their mutual interactions. If for given initial conditions, i.e. the values of

5.2 Molecular dynamics of simple fluids

$6N$ coordinates and momenta at time $t = 0$ can be specified, one can determine in principle the positions at a later time by means of $3N$ coupled differential equations of the type:

$$^\alpha\dot{\mathbf{p}} = \mu^\alpha\ddot{\mathbf{r}} = -\frac{\partial \mathscr{H}}{\partial^\alpha \mathbf{r}} = {}^\alpha\nabla U_N(^N\mathbf{r}) \qquad (5.2)$$

Thus, if $f(^N\mathbf{r},{}^N\mathbf{p})$ is a function of the $6N$-coordinates and momenta and f the corresponding thermodynamic quantity, the latter will be represented by the statistical average or:

$$\langle f \rangle_E = \langle f(^N\mathbf{r},{}^N\mathbf{p}) \rangle. \qquad (5.3)$$

In the classical theory such averages can be obtained in two ways. According to Boltzmann's theory the average $\langle f \rangle$ is taken as an average over time, i.e.:

$$\langle f \rangle_t = \lim_{\tau \to \infty} \frac{1}{\tau} \int_0^\tau f\{^N\mathbf{r}(t),{}^N\mathbf{p}(t)\}\,\mathrm{d}t. \qquad (5.4)$$

In this manner one can determine, for instance, the temperature of an isolated system from the time-averaged kinetic energy of the medium. For an isolated system the total energy is conserved, but the instantaneous kinetic energy will fluctuate with time (see subsequent consideration of the fluctuation theory). The magnitude of these fluctuations is related in such systems to the specific heat. Similarly one can determine the pressure in a fluid contained in a certain volume V by using Clausius' virial function which is expressed by:

$$\mathscr{V}(^N\mathbf{r}) = \sum_{\alpha=1}^{N} {}^\alpha\mathbf{r}\cdot{}^\alpha\mathbf{F} \qquad (5.5)$$

where $^N\mathbf{r}$ is the set of N position vectors and $^\alpha\mathbf{F}$ the total force acting on the αth molecule or particle. It can be readily shown that the time-average of this function $\langle \mathscr{V} \rangle_t = -3Nk_BT$ where k_B is Boltzmann's constant and T the temperature. The other way of averaging can be taken in terms of ensembles introduced by Gibbs. Thus, an ensemble in statistical mechanics is conventionally considered as a large number of replicas of the actual system, that may be infinite in general, which are all characterized by the same macroscopic parameters, but which occupy different micro-states with the particles having different positions and momenta. This description is based on the distribution of representative phase-points in a phase-space Γ, the corresponding probability density will be denoted here by $P^{(N)}$ so that:

$$P^{(N)}(^N\mathbf{r},{}^N\mathbf{p},t)\mathrm{d}^N\Gamma \equiv P^{(N)}(^N\mathbf{r},{}^N\mathbf{p},t)\mathrm{d}^N\mathbf{r}\,\mathrm{d}^N\mathbf{p} \qquad (5.6)$$

which represents the probability of finding the system at time t in a microscopic state characterized by a phase-point in Γ ($6N$-dimensional space). The infinitesimal volume in that space is:

$$d^N\Gamma = \prod_{\alpha=1}^{N} d^\alpha r\, d^\alpha p. \tag{5.7}$$

Under equilibrium conditions this probability density will be time independent and can be designated by $P_0^{(N)}(^N\mathbf{r}, {}^N\mathbf{p})$. The statistical average of any function $f(\mathbf{r}, \mathbf{p})$ in terms of the ensemble average is then obtained as:

$$\langle f \rangle_E = \iint f(^N\mathbf{r}, {}^N\mathbf{p}) P_0^{(N)}(^N\mathbf{r}, {}^N\mathbf{p}) d^N\mathbf{r}\, d^N\mathbf{p} \tag{5.8}$$

where $P_0^{(N)}$ is normalized, so that:

$$\iint P_0^{(N)}(^N\mathbf{r}, {}^N\mathbf{p}) d^N\mathbf{r}\, d^N\mathbf{p} = 1.$$

Thus the internal energy U can be expressed in terms of the average Hamiltonian of the system as follows:

$$U = \langle \mathcal{H} \rangle_E = \iint \mathcal{H}(^N\mathbf{r}, {}^N\mathbf{p}) P_0^{(N)}(^N\mathbf{r}, {}^N\mathbf{p}) d^N\mathbf{r}\, d^N\mathbf{p}. \tag{5.9}$$

The explicit form of the probability density function depends on the type of ensemble considered in the averaging procedure. It is well known from statistical mechanics that there are four types of ensembles, namely the canonical in which the pertaining macroscopic parameters are N, V, T (number of particles in the volume V, T is the temperature), the isothermal–isobaric ensemble with parameters N, P, T (P being the pressure), the grand canonical and the microcanonical ensembles. Of special interest in the present analysis is the canonical ensemble, which represents a collection of systems or subsystems that contain a fixed number of particles or molecules ($\alpha = 1, \ldots, N$) under the same conditions as the macroscopic volume and temperature. The most significant characteristic of this ensemble is its equilibrium probability density $P_0^{(N)}$, which on the assumption of identical particles becomes:

$$P_0^{(N)}(^N\mathbf{r}, {}^N\mathbf{p}) = \frac{1}{N!} \frac{1}{h^{3N}} \frac{1}{Q_N(V, T)} \exp[-\beta \mathcal{H}(^N\mathbf{r}, {}^N\mathbf{p})] \tag{5.10}$$

in which h is Planck's constant and the factor $N!$ originates from the assumption that all particles are indistinguishable. The latter assumption will be maintained in the following discussions.

The quantity $Q_N(V, T)$ is a normalizing factor known as the partition function and is defined by:

$$Q_N(V, T) = \frac{1}{N!} \frac{1}{h^{3N}} \iint \exp[-\beta \mathcal{H}(^N\mathbf{r}, {}^N\mathbf{p})] d^N\mathbf{r}\, d^N\mathbf{p}. \tag{5.11}$$

The multiplicative constant $(h^{3N})^{-1}$ in (5.10, 5.11) ensures that Q_N and $P_0^{(N)}$ are dimensionless and assume the corresponding expressions of quantum-statistical mechanics. Another quantity of interest in such an ensemble is the

5.2 Molecular dynamics of simple fluids   125

configuration integral Z_N defined by:

$$Z_N(V, T) = \int_V \exp[-\beta U_N(^N\mathbf{r})] d^N\mathbf{r} \tag{5.12}$$

in which the integration extends over each position vector $^\alpha\mathbf{r}$; ($\alpha = 1, 2, \ldots, N$) within the entire volume V containing the ensemble. The relation between Q_N and Z_N from thermodynamics is given by:

$$Q_N(V, T) = \frac{Z_N}{N! \lambda^{3N}} \tag{5.13}$$

where λ is the de Broglie thermal wavelength,

$$\text{i.e.} \quad \left[\frac{2\pi h^2}{^\alpha\mu}\right]^{\frac{1}{2}} \quad \text{and} \quad \beta = [k_B T]^{-1}$$

the inverse temperature of the fluid (see also Hirschfelder, Curtiss and Bird [66], Boon and Yip [156] and others). In connection with the concepts of time and ensemble averages in molecular dynamics of fluids a brief discussion on ergodic theorems relevant to the probabilistic mechanics approach is given in ref. [61, 36].

(ii) *Distribution functions*

For the description of the microstructure of fluids and as a measure of the correlations between the configurations of elements, the above-mentioned equilibrium probability density functions permit a rigorous analysis. It is, however, sufficient in calculating equilibrium properties in general to employ the lower order distribution or densities in the case of one-component monatomic fluids. Thus considering again the normalized canonical probability density for a fluid composed of $\alpha = 1, \ldots, N$ identical molecules, relation (5.10) can also be written as follows:

$$P_0^{(N)}(^N\mathbf{r},^N\mathbf{p}) = \left(\frac{\lambda}{h}\right)^{3/N} \exp[-\beta K_N(^N\mathbf{p})] \frac{\exp[-\beta U_N(^N\mathbf{r})]}{Z_N(V,T)} \tag{5.14}$$

in which $K_N(^N\mathbf{p})$ is the kinetic energy term in the Hamiltonian in (5.10). The probability density function can be factorized into $3N$-independent distributions (Maxwellian) for the components of the momenta and into a probability density for the coordinates. The latter does not in general separate due to correlations between positions of particles. Thus introducing the probability of simultaneously finding molecule 1 in the volume $d^1\mathbf{r}$ around $^1\mathbf{r}$

and molecule 2 in $d^2\mathbf{r}$ around $^2\mathbf{r}$, ... it can be stated that:

$$P_N^{(N)}(^1\mathbf{r}, \ldots, ^N\mathbf{r}) d^1\mathbf{r} \ldots d^N\mathbf{r} = \frac{1}{Z_N} \exp[-\beta U_N(^1\mathbf{r}, \ldots, ^N\mathbf{r})] d^1\mathbf{r} \ldots d^N\mathbf{r}. \quad (5.15)$$

This leads to the concept of an n-body probability density $P_N^{(n)}$ obtained from $P_N^{(N)}$ by integration over the coordinates of $(N - n)$ particles so that:

$$P_N^{(n)}(^1\mathbf{r}, \ldots, ^n\mathbf{r}) = \int \ldots \int P_N^{(N)}(^1\mathbf{r}, \ldots, ^N\mathbf{r}) d^{n+1}\mathbf{r} \ldots d^N\mathbf{r}. \quad (5.16)$$

Hence the average or mean of a function f of the coordinates of n-particles will be given by:

$$\langle f(^n\mathbf{r}) \rangle = \frac{1}{Z_N} \int \exp[-\beta U_N(^N\mathbf{r})] f(^n\mathbf{r}) d^N\mathbf{r} = \int P_N^{(n)}(^n\mathbf{r}') f(^n\mathbf{r}') d^n\mathbf{r}'. \quad (5.17)$$

Using the mutual distances between n-particles, expression (5.16) can also be written as:

$$P_N^{(n)}(^1\mathbf{r}, \ldots, ^n\mathbf{r}) = \langle \delta(^1\mathbf{r} - ^1\mathbf{r}') \ldots \delta(^n\mathbf{r} - ^n\mathbf{r}') \rangle. \quad (5.18)$$

In the limit $|^\alpha\mathbf{r} - ^\beta\mathbf{r}| \to \infty$ for all $1 \leq \alpha, \beta \leq n$, $P_N^{(n)}$ can be approximated by the product of the n-single particle probability densities so that:

$$P_N^{(n)}(^1\mathbf{r}, \ldots, ^n\mathbf{r}) \simeq P_N^{(1)}(^1\mathbf{r}) \ldots P_N^{(1)}(^n\mathbf{r}) \quad (5.19)$$

where in this limit the position of each α of $\alpha = 1, \ldots, n$ is independent of the positions of the $(n - 1)$ particles. Thus one can define the n-particle distribution functions in the following manner:

$$G_N^{(n)}(^1\mathbf{r}, \ldots, ^n\mathbf{r}) = \frac{P_N^{(n)}(^1\mathbf{r}, \ldots, ^n\mathbf{r})}{\prod_{\alpha=1}^{n} P_N^{(1)}(^\alpha\mathbf{r})}; \quad (5.20)$$

This shows that $G_N^{(n)}(^n\mathbf{r}) \to 1$ for all n, when the mutual distances between the n-particles increase $\to \infty$ (thermodynamic limit). Two distribution functions are most significant, i.e. the single-particle and the pair-distribution functions. For a homogeneous system and by translational invariance the single particle distribution function is given by:

$$P_N^{(1)}(^\alpha\mathbf{r}) = \frac{\int \ldots \int \exp[-\beta U_N(^N\mathbf{r})] d^1\mathbf{r} \ldots d^{\alpha-1}\mathbf{r} \cdot d^{\alpha+1}\mathbf{r} \ldots d^N\mathbf{r}}{\int \ldots \int \exp[-\beta U_N(^N\mathbf{r})] d^1\mathbf{r} d^2\mathbf{r} \ldots d^N\mathbf{r}}$$

$$= \frac{1}{U}; (\alpha = 1, \ldots, N) \quad (5.21)$$

so that in the n-particle case:

$$G_N^{(n)}(^n\mathbf{r}) = U^n P_N^{(n)}(^n\mathbf{r})$$

5.2 Molecular dynamics of simple fluids   127

and for two particles, the corresponding pair-distribution becomes:

$$G_N^{(2)}(^\alpha \mathbf{r}, ^\beta \mathbf{r}) = U^2 P_N^{(2)}(^\alpha \mathbf{r}, ^\beta \mathbf{r})$$

$$= \frac{1}{Z_N} U^2 \int \ldots \int \exp[-\beta U_N(^N\mathbf{r})] \quad (5.22)$$

$$\times d^1\mathbf{r} \ldots d^{\alpha-1}\mathbf{r}\, d^{\alpha+1}\mathbf{r} \ldots d^{\beta-1}\mathbf{r}\, d^{\beta+1}\mathbf{r} \ldots d^N\mathbf{r}. \quad (\alpha = 1, \ldots N)$$

If the system is also isotropic it is evident that:

$$G_N^{(2)}(^\alpha \mathbf{r}, ^\beta \mathbf{r}) = G(|^\alpha \mathbf{r} - ^\beta \mathbf{r}|). \quad (5.23)$$

The function $G(\mathbf{r})$ is usually referred to as the radial distribution function and becomes important in the molecular flow of interacting particles, which are subjected to central pair forces. In this case the potential energy for $\alpha = 1, \ldots, N$ particles and the Hamiltonian becomes:

$$U_N(^N\mathbf{r}) = \sum_{\alpha<\beta}^N \phi(^{\alpha\beta}\mathbf{r}) = \tfrac{1}{2} \sum_{\alpha\ne\beta}^N \phi(^{\alpha\beta}\mathbf{r}), \quad (5.24)$$

$$\mathscr{H} = \sum_{\alpha=1}^N \frac{^\alpha \mathbf{p}^2}{2\mu} + \tfrac{1}{2} \sum_{\alpha,\beta}{}' \phi(^{\alpha\beta}\mathbf{r}); \quad (\Sigma': \text{no summation over indices}).$$

(iii) *The concept of correlation functions*

1. *Dynamic variables.* For the analysis of microdynamical processes in solids and liquids, which are in or near the thermodynamic equilibrium, the notion of correlation functions is important. In particular the concept of time-correlation functions leads to a measure of the involved microscopic fluctuation of significant field variables. In fluids time-correlation functions are equally important for the single-particle motion and the collective mode of motion of a large number of particles or molecules. Such functions form also a link between the molecular or microscopic and the phenomenological description of conventional hydrodynamics. First, it is necessary to consider some definitions and the general formulation of such functions leading to linear response theory and fluctuation theory, which is then a natural transition to the probabilistic formulation of molecular fluid dynamics.

By considering a fluid system of $\alpha = 1, \ldots, N$ particles or molecules, any dynamic variable Y associated with the system will be a function of the generalized coordinates and momenta in the phase-space. It can be a scalar, vector or tensor valued quantity, the time evolution of which can be expressed in terms of an operator known as the Liouville operator. It is often convenient to use local dynamical variables, which generally may be expressed by:

$$Y(^\alpha\mathbf{r}, t) = \sum_{\alpha=1}^N {}^\alpha y(t) \delta(\mathbf{r} - ^\alpha\mathbf{r}(t)) \quad (5.25)$$

128   Probabilistic Mechanics of Fluids

in which $^\alpha y$ may be a physical quantity such as the mass $^\alpha\mu$ of the molecule, its linear or angular momentum and $^\alpha r(t)$ the position vector to its centre of mass (c.m.). In a quantum mechanical system $^\alpha y$ and $^\alpha \mathbf{r}$ in general do not commute so that (5.25) must be subjected to an appropriate symmetrization. In the following the spatial Fourier components of the variable $Y(^\alpha \mathbf{r},t)$ will be required, which are given by:

$$Y(\mathbf{k}, t) = \int Y(\mathbf{r}, t)e^{-i\mathbf{k}\cdot\mathbf{r}} = \sum_{\alpha=1}^{N} {}^\alpha y(t)e^{-i\mathbf{k}\cdot{}^\alpha\mathbf{r}(t)} \qquad (5.26)$$

where **k** is the wave vector.

If the dynamical variable is conserved, it will have to satisfy the continuity condition so that:

$$\frac{\partial Y(\mathbf{r}, t)}{\partial t} + \nabla \cdot \mathbf{j}(\mathbf{r}, t) = 0 \qquad (5.27)$$

where **j** represents the current related to the density of the variable Y. This relation also expresses the constancy of Y_{total},

which is equal to $\sum_{\alpha=1}^{N} {}^\alpha y$ in a continuous medium.

An important example in this context is the so-called number density ρ, i.e. ($^\alpha y = 1$) at any point in the medium:

$$\rho(\mathbf{r}, t) = \sum_{\alpha=1}^{N} \delta(\mathbf{r} - {}^\alpha \mathbf{r}(t)). \qquad (5.28)$$

Taking the Fourier transform of this density one obtains,

$$\rho(\mathbf{r}) = \sum_{\alpha=1}^{N} \delta(\mathbf{r} - {}^\alpha \mathbf{r})$$

and

$$\rho_\mathbf{k} = \int \exp[-i\mathbf{k}\cdot\mathbf{r}]\rho(\mathbf{r})d\mathbf{r} = \sum_{\alpha=1}^{N} \exp(-i\mathbf{k}\cdot{}^\alpha\mathbf{r}). \qquad (5.29)$$

The average particle density at an arbitrary point **r** can therefore be written as:

$$\left.\begin{aligned}\langle \rho(\mathbf{r}) \rangle &= \frac{N}{Z_N} \int \ldots \int \exp[-\beta U_N] d^2\mathbf{r} \ldots d^N\mathbf{r} \\ &= \rho^{(1)}(\mathbf{r}) \equiv \rho(\mathbf{r})\end{aligned}\right\} \qquad (5.30)$$

which is also the mean number density at any point in the continuous uniform medium. The associated current will be then:

$$\mathbf{j}_\rho(\mathbf{r}, t) = \sum_{\alpha=1}^{N} \frac{{}^\alpha\mathbf{p}^{(t)}}{{}^\alpha\mu} \delta(\mathbf{r} - {}^\alpha\mathbf{r}(t)) \qquad (5.31)$$

where $^\alpha\mu$ is the mass of the αth molecule and $^\alpha\mathbf{p}$ the linear momentum at its C.M. Taking the Fourier transform of a conserved dynamical variable, the continuity relation (5.27) can be stated in the form of:

$$\frac{\partial Y}{\partial t}(\mathbf{k}, t) + i\mathbf{k}\cdot\mathbf{j}(\mathbf{k}, t) = 0. \tag{5.32}$$

2. *Pair-correlation functions and structure factors.* In the molecular dynamics of a simple fluid the pair-correlation function is of utmost significance since it permits to establish relations for the physical properties of that fluid. In order to establish such a function in its time-independent and time-dependent form, it is necessary to introduce first the so-called equilibrium density-density correlation function. The latter can be defined in terms of the relative distance or separation between two molecules ($|^{\alpha\beta}\mathbf{d}| = |^\alpha\mathbf{r} - {}^\beta\mathbf{r}|$). This correlation function is defined by:

$$R_0(\mathbf{r}, \mathbf{r}') = \frac{1}{N} \langle [\rho(\mathbf{r}') - \langle\rho(\mathbf{r}')\rangle] [\rho(\mathbf{r}'+\mathbf{r}) - \langle\rho(\mathbf{r}'+\mathbf{r})\rangle] \rangle \tag{5.33}$$

showing the fluctuations around $\mathbf{r}$ of the number density $\rho(\mathbf{r})$. Since for a uniform system $R_0(\mathbf{r}, \mathbf{r}')$ depends only on the relative distance $|^{\alpha\beta}\mathbf{d}|$, $R_0(\mathbf{r})$ will be a function, that is independent of the choice of the origin so that:

$$R_0(\mathbf{r}) = \frac{1}{N} \int \langle \rho(\mathbf{r}')\rho(\mathbf{r}'+\mathbf{r}) \rangle \, d\mathbf{r}' - \rho \tag{5.34}$$

or equivalently

$$R_0(\mathbf{r}) = \frac{1}{N} \sum_{\alpha \neq \beta}^{N} \langle \delta(\mathbf{r} + {}^\alpha\mathbf{r} - {}^\beta\mathbf{r}) \rangle + \delta(\mathbf{r}) - \rho. \tag{5.35}$$

Using the pair-distribution function given in (5.23) one can write:

$$R_0(\mathbf{r}) = \rho G^{(2)}(\mathbf{r}) + \delta(\mathbf{r}) - \rho = \rho R^{(2)}(\mathbf{r}) + \delta(\mathbf{r}) \tag{5.36}$$

in which $R^{(2)}(\mathbf{r})$ denotes the equilibrium pair-correlation function or $^{\alpha\beta}R$ in terms of the relative distance $|^{\alpha\beta}\mathbf{d}|$. It is defined as follows:

$$^{\alpha\beta}R \equiv R^{(2)}(^{\alpha\beta}\mathbf{d}) = G^{(2)}(^{\alpha\beta}\mathbf{d}) - 1. \tag{5.37}$$

This relation indicates that for a uniform system where $G^{(2)}(\mathbf{r})$ has an asymptotic behaviour, $R^{(2)}(\mathbf{r}) \to 0$, when $|\mathbf{r}| \to \infty$. Hence $|\mathbf{r}|$ is a significant quantity in the flow of molecular fluids with interactions. It will be further discussed in the probabilistic mechanics of discrete fluids in which $|^{\alpha\beta}\mathbf{d}|$ is seen as a metric in the corresponding configuration space as a subspace of the general probabilistic function space.

In the classical theory the equilibrium time-correlation function pertaining to two dynamical variables Y, Z is important. It is defined by:

$$R_{YZ}(t', t'') = \langle Y(t')Z(t'') \rangle \tag{5.38}$$

where the $\langle \cdot \rangle$ designates either an ensemble average over initial conditions or an average over time as discussed in section 5.1. It is always assumed that the system under consideration is ergodic in the sense discussed in reference [61]. In the present context, considering the time-average of the above variables one obtains:

$$\langle Y(t')Z(t'') \rangle = \lim_{T \to \infty} \frac{1}{T} \int_0^T Y(t'+s)Z(t''+s) \mathrm{d}s. \tag{5.39}$$

It is apparent that, if the variables are also subjected to a spatial variation (5.25), the time-correlation function will have to be formulated both in time and space and is then given in its non-local form by:

$$R_{YZ}(\mathbf{r'}, \mathbf{r''}; t', t'') = \langle Y(\mathbf{r'}, t') Z(\mathbf{r''}, t'') \rangle. \tag{5.40}$$

Considering the Fourier components of these variables which in general are complex quantities the time-correlation function can be defined in the form of:

$$\left. \begin{array}{l} R_{YZ}(\mathbf{k'}, \mathbf{k''}; t', t'') = \langle Y(\mathbf{k'}, t') Z^*(\mathbf{k''}, t'') \rangle \\ \qquad\qquad\qquad\quad = \langle Y(\mathbf{k'}, t') Z(-\mathbf{k''}, t'') \rangle \end{array} \right\} \tag{5.41}$$

since the quantities $^\alpha y$ are real in (5.25).

In the special case, where a dynamical variable is correlated with itself the time-correlation function is known as the auto-correlation function discussed in Chapter 1 and hence is given by:

$$R_{yy}(t', t'') = \langle Y(t'), Y(t'') \rangle. \tag{5.42}$$

Considering in particular the auto-correlation function of the Fourier components of the particle density, which defines a so-called structure factor $S(\mathbf{k})$ of the fluid, i.e.:

$$S(\mathbf{k}) = \frac{1}{N} \langle \rho_k \rho_{-k} \rangle. \tag{5.43}$$

The structure factor $S(\mathbf{k})$ can also be obtained from the Fourier transform of the pair-distribution function. Thus by using relations (5.43) and (5.29), it is seen that:

$$S(\mathbf{k}) = \frac{1}{N} \sum_{\alpha=1}^{N} \sum_{\beta=1}^{N} \exp[-i\mathbf{k} \cdot {}^\alpha \mathbf{r}] \exp[i\mathbf{k} \cdot {}^\beta \mathbf{r}]. \tag{5.44}$$

It can be readily shown that this leads to the form:

$$S(\mathbf{k}) = 1 + \rho \int \exp(-i\mathbf{k} \cdot \mathbf{r}) G(\mathbf{r}) \mathrm{d}\mathbf{r} \tag{5.45}$$

where $G(\mathbf{r})$ is the radial distribution function given in (5.23). Using the Fourier transform of $S(\mathbf{k})$ this distribution function will be given by:

$$\rho G(\mathbf{r}) = \frac{1}{(2\pi)^3} \int [S(\mathbf{k}) - 1] \exp(i\mathbf{k} \cdot \mathbf{r}) \mathrm{d}\mathbf{k}. \tag{5.46}$$

5.2 Molecular dynamics of simple fluids

This structure factor $S(\mathbf{k})$ is also referred to as static structure factor. It can be determined directly from scattering experiments (see also Lovesey [157], Dore et al. [158, 159] and others). It is to be noted that in an isotropic fluid where $G(\mathbf{r}) \equiv G(r)$ and $S(\mathbf{k}) \equiv S(k)$, $G(\mathbf{r})$ describes the average distribution of the molecular separation in the fluid. It is of interest, however, to obtain a dynamic structure factor, which is a function of both the wave vector $\mathbf{k}$ and the frequency ω of the wave motion. For this purpose one has to consider the correlation in space and time as well as a scattering function.

Without giving a discussion on scattering theory, the dynamic structure factor can be defined in terms of the wave vector $\mathbf{k}$ and the frequency ω as follows:

$$\left.\begin{aligned} S(\mathbf{k}, \omega) &= \frac{1}{2\pi N} \int_{-\infty}^{\infty} \exp(i\omega t) \langle \rho_\mathbf{k}(t) \rho_{-k} \rangle \, dt \\ &= \frac{1}{2\pi} \int_{-\infty}^{\infty} \exp(i\omega t) F(\mathbf{k}, t) \, dt \end{aligned}\right\} \quad (5.47)$$

where $F(\mathbf{k}, t)$ in the integrand is called an intermediate scattering function.

The latter can also be defined in terms of the time-dependent Fourier components of the density given earlier so that:

$$F(\mathbf{k}, t) = \frac{1}{N} \langle \rho_\mathbf{k}(t) \rho_{-k} \rangle \quad (5.48)$$

which upon comparison with (5.43) shows that: $F(\mathbf{k}, 0) = S(\mathbf{k})$ or the static structure factor.

Although no calculation of this structure factor $S(\mathbf{k})$ is intended in this text some of the basic properties of this function should be briefly summarized. There are three functions that are conventionally used to describe spatial correlations in fluids. Thus, apart from $S(\mathbf{k})$ one has the pair-correlation function which is related to the structure factor by:

$$S(k) = 1 + \rho R^{(2)}(k) \quad (5.49)$$

since $F(k, 0) = 1 + \rho \int d^2 r \, e^{i\mathbf{k}\cdot\mathbf{r}} [R_0(r) - 1] = S(k)$.

It is seen that $R^{(2)}(r)$ is the Fourier transform of $[R_0(r) - 1]$. Usually a direct correlation function is also considered of the form:

$$\rho C(k) = \frac{S(k) - 1}{S(k)} \quad (5.50)$$

or
$$R^{(2)}(k) = R(k) + \rho R(k) R^{(2)}(k) \quad (5.51)$$

which is known as the Ornstein–Zernicke relation (see Hansen [72, 155]). It is a fundamental relation in the theory of pair-distribution functions. The behaviour of the structure factor $S(k)$ is generally analogous to that of the pair-distribution function. Thus if k is large, the asymptotic value of $S(k)$ approaches unity, hence indicating the vanishing pair-correlation at short wavelength. In the other limit, however, i.e. for small k one obtains the known compressibility relation linking the structure factor to the isothermal compressibility of a dilute fluid (see amongst others, Egelstaff [160]).

It can be seen from the brief discussion above that the particle density or the number density ρ is a characteristic of the most general interest in molecular dynamics. Experimental work directed towards assessing the various frequency domains (k, ω) of a discrete fluid can be grouped according to two significant parameters, i.e. a characteristic frequency ω_c and a characteristic wave number k_c. The first corresponds to the reciprocal collision time $(\tau_c \sim 10^{12}$ to 10^{13} sec$^{-1})$ and the second to the reciprocal of the intermolecular distance $(^{\alpha\beta}d_c^{-1} \sim 10^8$ cm$^{-1})$. The domain bounded by the values $\omega_c \sim \tau_c^{-1}$, $k_c \sim {}^{\alpha\beta}d_c^{-1}$ is the region of short wavelength and high frequencies, where molecular dynamics applies.

Apart from the experimental work carried out for this domain of the dynamical behaviour of simple fluids, it should be mentioned that more recently computer-simulation techniques are used instead of such experiments. The most frequently employed method in this context is the Monte-Carlo technique to be discussed briefly in section 5.5 (see also Metropolis *et al.* [161] for the basic formulation and more recently ref. [162]).

5.3 RESPONSE AND FLUCTUATION THEORY

In the foregoing sections the significance of the concept of correlation functions in the formulation of molecular processes in fluids at or near the thermodynamic equilibrium has been stressed. From a statistical point of view these functions relate directly to the microscopic fluctuations of the relevant quantities in the fluid system. It is equally of interest to study the response of a fluid, when it is perturbed by an external field. This is usually carried out in terms of the time-dependent correlation functions, which then corresponds to a linearization of the hydrodynamic relations in the limit, i.e. if the physical quantities vary slowly in space and time. Hence by using certain characteristics of the obtained response functions one can derive correlation functions, which hold also for high frequencies and short wavelengths. This formalism in the theory of transport processes has been often used and is discussed amongst others by Martin [163], Martin and Yip [164], Chung and Yip [165], Hansen and McDonald [72].

(i) *The response theory formalism*

It is necessary first to show that the linear response of a fluid system to an external perturbation is related to fluctuations in the representation of the

5.3 Response and fluctuation theory

system at equilibrium. Thus considering the Hamiltonian of a system in the presence of an external field, then:

$$\mathcal{H} = \mathcal{H}_0 + \mathcal{H}' \qquad (5.52)$$

where $\mathcal{H}_0$ is the Hamiltonian of the unperturbed system and $\mathcal{H}'$ its perturbation, which is in general time-dependent. If $F(\mathbf{r}', t)$ denotes the time-dependent external field that may be of the scalar, vector or tensorial type and $Y(\mathbf{r}', t)$ a dynamical variable coupled to it, then the Hamiltonian $\mathcal{H}'$ will be of the form:

$$\mathcal{H}'(t) = -\int Y(\mathbf{r}')F(\mathbf{r}', t)\mathrm{d}^3 r' \qquad (5.53)$$

where the minus sign is a convention. To show the linear response in terms of the fluctuations of the properties of the fluid system, one may consider a variable Z and its changes due to $F(t)$ coupled to the variable Y (see Kubo [166]). In this case without specifying the particular form of coupling $\mathcal{H}'(t) = -YF(t)$. For a fluid system consisting of N-particles or molecules, the N-particle probability distribution will change and can be expressed in terms of the Liouville operators $\mathcal{L}$ and $\mathcal{L}_0$, where the latter refers to the unperturbed Hamiltonian $\mathcal{H}_0$. Hence,

$$\left. \begin{aligned} \frac{\partial P^{(N)}(t)}{\partial t} &= -i\mathcal{L}P^{(N)}(t) = \{\mathcal{H}, P^{(N)}(t)\} \\ &= -i\mathcal{L}_0 P^{(N)}(t) - \{Y, P^{(N)}(t)\}F(t). \end{aligned} \right\} \qquad (5.54)$$

Assuming that the external field $F(t)$ is applied, when the system is in its thermal equilibrium, results in the probability distribution:

$$P_0^{(N)} = C\exp(-\beta\mathcal{H}_0) \qquad (5.55)$$

where the constant C here is equal to the factor in front of the exponent in (5.10). Since the linear response of the system only is considered here, one can express $P^{(N)}(t)$ as follows:

$$P^{(N)}(t) = P_0^{(N)} + \Delta P^{(N)}(t) \qquad (5.56)$$

and hence for the case of small perturbations (5.54) can be expressed by:

$$\frac{\partial}{\partial t}\Delta P^{(N)}(t) = -i\mathcal{L}_0 \Delta P^{(N)}(t) - \{Y, P_0^{(N)}\}F(t). \qquad (5.57)$$

Integrating the above form gives:

$$\Delta P^{(N)}(t) = -\int_{-\infty}^{t} \exp[-i(t-s)\mathcal{L}_0]\{Y, P_0^{(N)}\}F(s)\mathrm{d}s. \qquad (5.58)$$

Taking now the average change in the variable Z, i.e. $\langle \Delta Z(t) \rangle$ caused by the change in the distribution function, then:

$$\langle \Delta Z(t) \rangle = - \iint d^N\mathbf{r}\, d^N\mathbf{p} \int_{-\infty}^{t} \{Y, P_0^{(N)}\} Z[t-s] F(s)\, ds \tag{5.59}$$

where $Z(t) = \exp[i\mathscr{L}_0 t]Z$.
This results from considerations of the right-hand side of (5.58) considered as an inner product of the quantities $\{Y, P_0^{(N)}\}$ and Z gives also:

$$- \iint d^N\mathbf{r}\, d^N\mathbf{p} \int_{-\infty}^{t} ds \{Y, P_0^{(N)}\} \exp[i(t-s)\mathscr{L}_0] Z F(s).$$

By introducing the so-called after-effect function $\phi_{zy}(t)$ defined by:

$$\left.\begin{array}{ll} \phi_{zy}(t) = - \iint \{Y, P_0^{(N)}\} Z(t) d^N\mathbf{r}\, d^N\mathbf{p}, & t \geq 0 \\ = 0, & t < 0 \end{array}\right\} \tag{5.60}$$

the average change of $Z(t)$ will be given by:

$$\langle \Delta Z(t) \rangle = \int_{-\infty}^{t} \phi_{zy}[t-s] F(s)\, ds. \tag{5.61}$$

Since ϕ_{zy} vanishes for $t < 0$ the upper limit in (5.61) can be extended to $+\infty$. The linear response to an externally applied field $F(t)$ can be regarded in general as a superposition of monochromatic components of the form $F_0 \exp(-i\omega t)$, where ω is the same wave frequency as before.

To ensure that the perturbations vanish in the limit when $t \to -\infty$, one can include the factor $\exp(\varepsilon t); \varepsilon \to 0^+$ in the formula for F_0. Since a linear response is assumed it is sufficient to consider one component only so that the response to a real field can be written as:

$$\langle \Delta Z(t) \rangle = \operatorname{Re} \chi_{ZY}(\omega) F_0 \exp(-i\omega t) \tag{5.62}$$

in which $\chi_{ZY}(\omega)$ is a complex dynamic function also called the susceptibility function. Its Fourier transform is given by:

$$\chi_{ZY}(\omega) = \chi'_{ZY}(\omega) + \chi''_{ZY}(\omega) \tag{5.63}$$

and represents the $\lim_{\varepsilon \to 0^+} \int_0^\infty \phi_{ZY}(t) \exp[i(\omega + i\varepsilon)t]\, dt$ where χ'_{ZY} is the real and χ''_{ZY} the imaginary parts, respectively.

Using the time correlation function $R_{YZ}(t)$ from before and its Fourier transform $\hat{R}_{ZY}(\omega)$ the latter will relate to the susceptibility function as follows:

$$\chi_{ZY}(\omega) = \beta[\langle ZY \rangle + i\omega \hat{R}_{ZY}(\omega)]. \tag{5.64}$$

5.3 Response and fluctuation theory

In the case when the variables $Z = Y$ the corresponding autocorrelation function in terms of frequencies can be given by:

$$\left. \begin{array}{l} R_{YY}(\omega) = \dfrac{1}{2\pi} \displaystyle\int_{-\infty}^{\infty} R_{YY}(t) \exp(i\omega t) \, dt \\[2ex] \qquad\quad = \dfrac{K_B T}{\pi \omega} \chi''_{YY}(\omega). \end{array} \right\} \qquad (5.65)$$

This relation is one representation of the so-called fluctuation–dissipation theorem (see Green, Kubo [168, 167]). The term dissipation relates to the fact, that energy is absorbed from the applied external field and dissipated in the form of heat, which is proportional to the function $\chi''_{ZY}(\omega)$. Similar forms to equation (5.65) can be obtained that relate transport coefficients to the integral of time-correlation functions. The approach using a linear response theory is not always so simple, since in many cases the derivation of transport coefficients may be very complex. A review on such problems is due to Zwanzig [169]. The application of correlation functions and the notion of a generalized hydrodynamics with regard to transport properties of simple fluids is also dealt with by Kadanoff and Martin [170], Chung and Yip [165], Hansen and McDonald [72] and others.

(ii) *Fluctuations of a dynamic variable*

In the foregoing sections mean values of dynamic variables and the relevant correlations functions have been considered. Most descriptions of transport phenomena are based on the theory of Brownian motion, although this motion in its classical form was concerned with the problem of self-diffusion. The interpretation of the Brownian motion due to Langevin is, however, of interest here, since it permits the introduction of probabilistic concepts to the dynamics of discrete fluids in a natural way. It will be briefly discussed in the following. It may be instructive to consider first the time evolution of fluctuations of a dynamical variable $Y(t)$. Such fluctuations are given in terms of the mean values by:

$$\Delta Y(t) = Y(t) - \langle Y(t) \rangle. \qquad (5.66)$$

The time evolution of $Y(t)$ in general can be expressed by a stochastic differential equation of the Langevin type as follows:

$$\frac{d}{dt} Y(t) + \gamma Y(t) = \tilde{f}(t) \qquad (5.67)$$

where γ is a coefficient related to the dissipation effect occurring during the motion and $\tilde{f}(t)$ a random fluctuating force with mean value zero. This differential equation is in fact Langevin's phenomenological interpretation of the Brownian motion of a particle. The latter was originally seen as the motion of a free particle of mass μ moving with a small velocity in a fluid. In this sense the motion can be described in the form of a linear friction law so that:

$$\mu \frac{dv}{dt} + \gamma v = 0 \tag{5.68}$$

where γ here merely indicates a friction constant. The particle moves to rest exponentially with a damping constant γ/μ. More specifically the Brownian system is considered to consist of heavy particles (mass μ) suspended in a fluid of light particles of mass m ($m \ll \mu$) and where the Brownian particles do not interact with each other, but only with the light particles. Thus the motion can be expressed by:

$$\mu \frac{d}{dt} v + \gamma v = \tilde{f}(t) \tag{5.69}$$

which is of identical form to (5.67). Hence Langevin's interpretation of (5.68) is that, it represents an average relation, and which is extended by adding a random fluctuating force $\tilde{f}(t)$. The latter is meant to account for the effect of molecular interactions (collisions), which is not contained in the linear friction term of (5.68). Conceptually one separates therefore a systemic effect of molecular interaction and a random part expressed by the random force $\tilde{f}(t)$. To obtain an explicit form of this force, it is usually assumed, that the fluctuations of $\tilde{f}(t)$ are of the Gaussian form and represented by a Gaussian random process with mean value zero. As shown in Chapter 1 (eqn. (1.135)) this process has a joint probability distribution given by:

$$P\{x_1 t_1; \ldots; x_n t_n\} = \exp\left[-\frac{1}{2}\sum_{i,k}^{n} Y_{ik} x_i x_k\right] \frac{\det \mathbf{Y}^{1/2}}{(2\pi)^{n/2}} \tag{5.70}$$

where the matrix $\mathbf{Y}$ or $Y_{ik} = Y_{ik}(t)$ has the det $\mathbf{Y} = ((Y_{ik}))$. If the covariance matrix is denoted by $\mathbf{C} = ((\rho_{ik}))$ with the identity $\mathbf{CY} = \mathbf{I}$, then the covariance function becomes:

$$\rho_{kk} = \langle x^2(t_k) \rangle \quad \text{and} \quad \rho_{ik} = \langle x(t_i) x(t_k) \rangle. \tag{5.71}$$

For a stationary process ρ_{kk} is independent of k and is a constant so that: $\rho_{kk} = \langle x^2(t_k) \rangle = \sigma^2$ (variance), ρ_{ik} depends only on the time difference $t_i - t_k$: $\rho_{ik} = \rho(|t_i - t_k|)$.

The function ρ is non-negative definite and hence this process is completely determined by $\rho(t)$. It has been shown by Doob [33] that a stationary Gaussian process is also Markovian, if and only if the covariance function is of the form:

$$\rho(t) = e^{-\gamma |t|}. \tag{5.72}$$

5.3 Response and fluctuation theory

Returning to the Langevin type stochastic differential equation (5.67) and making the assumption that the fluctuating force has the properties which give a solution of this equation corresponding to a stationary Gaussian Markov process, it then will be of the form:

$$Y(t) = Y_0 e^{-\gamma t} + \int_0^t e^{-\gamma(t-\tau)} \tilde{f}(\tau) \, d\tau. \tag{5.73}$$

Then

$$\langle Y(t) \rangle = Y_0 e^{-\gamma t}, \quad \text{if} \quad Y(0) \equiv Y_0 \quad \text{and} \quad \langle \tilde{f}(t) \rangle = 0. \tag{5.74}$$

The assumption of stationarity of the process leads to a variance of the following form:

$$\langle [Y(t) - Y_0 e^{-\gamma t}]^2 \rangle = \sigma^2 [1 - e^{-2\gamma t}]. \tag{5.75}$$

Since the averaging has to be carried out twice, the second time over an ensemble of initial values $Y(0)$, which has a Gaussian distribution function given by:

$$P(Y_0 \leq y) = \frac{1}{\sigma \sqrt{2\pi}} \int_{-\infty}^{y} \exp\left[-\frac{(y' - Y_0)^2}{2\sigma^2}\right] dy' \tag{5.76}$$

the variance will be obtained as:

$$\langle [Y(t) - Y_0 e^{-\gamma t}]^2 \rangle = \int_0^t \int_0^t e^{-\gamma(t-\tau_1)} e^{-\gamma(t-\tau_2)} \langle \tilde{f}(\tau_1) \tilde{f}(\tau_2) \rangle \, d\tau_1 d\tau_2 \tag{5.77}$$

and where it is readily verified that for $Y(t)$ to be stationary, the random force $\tilde{f}(t)$ must also be stationary. It will be of a Gaussian form, if the covariance function is singular and proportional to $\delta(t - \tau)$. Such a random process is known as white noise, since its power spectrum is flat. By considering the Langevin equation for the Brownian particle (eqn. (5.69)) and assigning the specific properties to the fluctuating force $\tilde{f}(t)$ according to a Gaussian process:

$$\left. \begin{array}{l} \langle \tilde{f}(t) \rangle = 0, \\ \langle \tilde{f}(\tau_1) \tilde{f}(\tau_2) \rangle = c \delta(\tau_2 - \tau_1) \end{array} \right\} \tag{5.78}$$

in which c is a constant, an explicit form for the random force can be obtained. More specifically, the second line of (5.78) states that $\tilde{f}(t)$ is uncorrelated at different times. Hence its autocorrelation function can also be expressed by:

$$\langle \tilde{f}(0) \tilde{f}(t) \rangle = c \delta(t) \quad \text{(a)}$$

with the spectrum

$$\langle |\tilde{f}(\omega)|^2 \rangle = 2c. \quad \text{(b)} \tag{5.79}$$

Thus expressing the solution of the general relation for $Y(t)$ (eqn. (5.67)) in its time Fourier transform, one gets:

$$Y(\omega) = \frac{\tilde{f}(\omega)}{i\omega + \gamma} \tag{5.80}$$

which gives the power spectrum of $Y(t)$ as follows:

$$\langle |Y(\omega)|^2 \rangle = \frac{\langle |\tilde{f}(\omega)|^2 \rangle}{\omega^2 + \gamma^2} = \frac{2c}{\omega^2 + \gamma^2}. \tag{5.81}$$

On the basis of stationarity of the random process, one can define a static mean square fluctuation of the variable Y by:

$$\langle |Y|^2 \rangle = \frac{1}{2\pi} \int_{-\infty}^{\infty} \langle |Y(\omega)|^2 \rangle \, d\omega \tag{5.82}$$

so that from (5.80) and (5.81) the mean square fluctuation is related to the dissipation coefficient γ by:

$$\langle |Y|^2 \rangle = \gamma^{-1} c. \tag{5.83}$$

Eliminating c by substituting (5.83) into (5.79(b)) gives then:

$$\langle |\tilde{f}(\omega)|^2 \rangle = 2\gamma \langle |Y|^2 \rangle \tag{5.84}$$

stating the well-known fluctuation–dissipation theorem, which expresses the balance between the fluctuations $\langle |Y|^2 \rangle$ and the dissipation coefficient γ. By taking the value of c from (5.83) and substituting it into (5.79(a)) shows upon integration over time that:

$$\gamma = \langle |Y|^2 \rangle^{-1} \cdot \int_{-\infty}^{\infty} \langle \tilde{f}(0)\tilde{f}(t) \rangle \, dt. \tag{5.85}$$

Hence the phenomenological dissipation coefficient is given by the time integral of the stochastic force autocorrelation function. It is apparent, that the fluctuation–dissipation theorem applies to the long-time limit only. Thus equation (5.67) will be correctly expressing the behaviour of $Y(t)$ in the asymptotic limit, i.e. for $t \to \infty$ and if it is assumed that the dissipation coefficient γ is constant. This can be illustrated by taking $Y(t)$ to be the velocity of the Brownian particle $v(t)$ at time t given that $v(0) = v_0$ (see, for instance, Wang and Uhlenbeck [171]). Considering the conditional probability $P(v_0|v;t)$, it will attain in the limit for $t \to \infty$, the distribution in the form of:

$$P(v) = \sqrt{\frac{\gamma}{(2\pi c)}} e^{-\gamma v^2/2c} \tag{5.86}$$

5.3 Response and fluctuation theory

in which the initial value v_0 will disappear in this limit. However, from a statistical mechanics point of view, this distribution should be a Maxwellian distribution on grounds of the equipartition theorem so that it becomes also:

$$P(v) = \sqrt{\frac{\mu}{(2\pi kT)}}\, e^{-\mu v^2/2kT}. \tag{5.87}$$

Obviously the above relations can only be consistent, if $\gamma/c = \mu/kT$ or $c = kT\gamma/\mu$, which expresses the relation between the constant c and the dissipation coefficient γ. It can be readily shown by taking $v(t)$ as a Gaussian process with an exponential covariance function that for times $t \gg \mu/\gamma$, one obtains the well-known Einstein diffusion relation, i.e.:

$$D = kT/\gamma$$

in which D is the diffusion constant. Thus for long times the dissipation effects are proportional to $1/\gamma$. This is not the case, however, when $t \ll \gamma^{-1}$ since memory effects must be considered. This evidently requires a generalization of Langevin's equation. A detailed discussion on this concept and the application of memory functions in molecular dynamics of fluids is given, for instance, by Boon and Yip [156].

With regard to experimental work some recent results concerning the motion of Brownian particles with interactions are of interest. Thus using intensity fluctuation techniques (spectroscopy) Pusey [172], Ackerson [173], Berne [174] have shown that there is an important time scale in a dispersed system, which is of the order of the collision time τ_{coll}. It is the time taken by a particle to move a fraction of the mean particle spacing. For times shorter than τ_{coll} the dominant influence in the particle motion is a strongly fluctuating force caused by the collisions with the solvent molecules. This force induces the usual free particle Brownian motion. If, however, the time is greater than τ_{coll}, the weaker but longer-lasting particle interaction force becomes more significant, which leads to an interacting motion. It is this type of motion which will be discussed in the subsequent sections of the probabilistic mechanics formulation of discrete fluid flow. Finally, considerations of fluctuation phenomena far from equilibrium are of considerable importance since they relate in general to the transients and stability of the system, but will not be persued here.

(iii) *The Fokker–Planck equation*

The significance of the Kolmogorov differential equation in the stochastic mechanics of solids has been stressed in the foregoing chapter. Another form of this relation known as the forward Kolmogorov equation is called the Fokker–Planck equation which will be considered in this section. It is one of the basic relations required in the analysis of diffusion problems, fluctuation and scaling theories employed in the dynamics of fluids. There are various ways of obtaining this relation (see, for instance, Bharucha-Reid [121], Prohorov and Rozanov [19], Kac [63], Kolmogorov [2], Feller [28]). Thus considering first the backward Kolmogorov equation the method given by Kolmogorov will be used for the derivation of the forward equation.

Thus considering a continuous Markov process $\{x(t), t \geq 0\}$ characterized by a transition function in a given subspace $X \subset \mathscr{X}$ and following the general discussion given in section 2.2 (ii) of Chapter 2, this transition function can also be expressed by:

$$P(s, \xi; t, \eta) = \mathscr{P}\{x(t) < \eta | x(s) = \xi\}; \quad t > s \tag{5.88}$$

where $P(s, \xi; t, \eta)$ is a continuous function of t for a fixed s and ξ. It is also a conditional distribution function in η satisfying the conditions:

$$\lim_{\eta \to -\infty} P(s, \xi; t, \eta) = 0 \quad \text{and} \quad \lim_{\eta \to \infty} P(s, \xi; t, \eta) = 1. \tag{5.89}$$

It is assumed that this transition function has also a density function given by:

$$p(s, \xi; t, \eta) = \frac{\partial P(s, \xi; t, \eta)}{\partial \eta}. \tag{5.90}$$

To obtain the backward equation, it is assumed that the transition function (5.88) satisfies for a small time interval the following conditions:

$$\left.\begin{aligned}
&\text{(i)} \lim_{\Delta t \to 0} \frac{1}{\Delta t} \int_{|\eta - \xi| > \varepsilon} P(s, \xi; s + \Delta t, d\eta) = 0, \\
&\text{(ii)} \lim_{\Delta t \to 0} \frac{1}{\Delta t} \int_{|\eta - \xi| \leq \varepsilon} (\eta - \xi) P(s, \xi; s + \Delta t, d\eta) = a(s, \xi), \\
&\text{(iii)} \lim_{\Delta t \to 0} \frac{1}{\Delta t} \int_{|\eta - \xi| \leq \varepsilon} (\eta - \xi)^2 P(s, \xi; s + \Delta t, d\eta) = d^2(s, \xi)
\end{aligned}\right\} \tag{5.91}$$

in which $\varepsilon > 0$ is a positive number and $a(s, \xi)$ called the drift coefficient that characterizes on the average the evolution of $x(s)$ in the small time interval from $s \to s + \Delta t$ under the condition that $x(s) = \xi$. The quantity $d(s, \xi)$ determines the mean square deviation of $x(s)$ from its expected value and is called the diffusion coefficient. Assuming that the limit conditions are satisfied uniformly with respect to s, one can introduce a test function, which is bounded and continuous of the form:

$$\phi(s, \xi) = \int \phi(\eta) P(s, \xi; t, d\eta); \quad t > s. \tag{5.92}$$

This function on the assumption that the derivatives

$$\frac{\partial}{\partial \xi} \phi(s, \xi) \quad \text{and} \quad \frac{\partial^2}{\partial \xi^2} \phi(s, \xi)$$

5.3 Response and fluctuation theory   141

are bounded and continuous will have a derivative with respect to s, i.e. $\dfrac{\partial}{\partial s}\phi(s,\xi)$ satisfying the following differential equation:

$$\left.\begin{aligned}\frac{\partial}{\partial s}\phi(s,\xi) &= -a(s,\xi)\frac{\partial}{\partial \xi}\phi(s,\xi) - \frac{1}{2}d^2(s,\xi)\frac{\partial^2}{\partial \xi^2}\phi(s,\xi)\\ \text{with a limit condition:} & \\ \lim_{s\to t}\phi(s,\xi) &= \phi(\xi).\end{aligned}\right\} \quad (5.93)$$

Using the transition probability density defined by (5.90) and assuming that it is continuous with respect to s together with its first and second derivatives with respect to ξ, then it is a fundamental solution of the following differential equation:

$$\frac{\partial}{\partial s}p(s,\xi;t,\eta) = -a(s,\xi)\frac{\partial}{\partial \xi}p(s,\xi;t,\eta) - \frac{1}{2}d^2(s,\xi)p(s,\xi;t,\eta) \quad (5.94)$$

which is referred to as the backward Kolmogorov equation. To obtain the forward equation, which is the adjoint of the above relation, it is also assumed that the transition probability density (5.90) exists and has a derivative

$$\frac{\partial}{\partial s}p(s,\xi;t,\eta),$$

which is continuous with respect to s and η. It is further assumed that the following partial derivatives exist:

$$\frac{\partial}{\partial \eta}[a(t,\eta)p(s,\xi;t,\eta)] \quad \text{and} \quad \frac{\partial^2}{\partial \eta^2}[d^2(t,\eta)p(s,\xi;t,\eta)].$$

It has been shown by Bharucha-Reid [121] that by introducing first a non-negative continuous function satisfying:

$$\left.\begin{aligned}\phi(\eta) &= 0 \quad \text{for} \quad \eta < \eta_1 \quad \text{and} \quad \eta < \eta_2, \quad \eta_1 < \eta_2,\\ \phi(\eta_1) &= \phi(\eta_2) = \phi'(\eta_1) = \phi'(\eta_2) = \phi''(\eta_1) = \phi''(\eta_2) = 0\end{aligned}\right\} \quad (5.95)$$

as well as a second function $\phi(\zeta)$, which is bounded (conditions 5.95) and expanding it about η, leads to the result that $p(s,\xi;t,\eta)$ represents a fundamental solution of the differential equation:

$$\left.\begin{aligned}\frac{\partial}{\partial s}p(s,\xi;t,\eta) &= -\frac{\partial}{\partial \eta}[\alpha(t,\eta)p(s,\xi;t,\eta]\\ &\quad + \frac{1}{2}\frac{\partial^2}{\partial \eta^2}[d^2(t,\eta)p(s,\xi;t,\eta)]\end{aligned}\right\} \quad (5.96)$$

which is the forward Kolmogorov differential equation or the Fokker–Planck equation. Most of the well-known diffusion processes are special cases of this relation. The scalar representation given above can be readily extended to vector-valued Markov processes that consist of several independent random functions $a_i(t)$, i.e. $\mathbf{a}_N(t) = \{a_1(t), \ldots, a_N(t)\}$. The probability density satisfies then a multi-dimensional Fokker–Planck equation of the form:

$$\frac{\partial p}{\partial t} = -\sum_{i=1}^{N} \frac{\partial}{\partial a_i}[A_i(\mathbf{a}_N)p] + \sum_{i,j=1}^{N} \frac{\partial^2}{\partial a_i \partial a_j}[B_{ij}(\mathbf{a}_N)p]. \tag{5.97}$$

The existence and uniqueness of solutions of the Kolmogorov differential equations have been studied extensively. The theorems concerning these solutions in terms of the semi-group theory are given by Feller [28] (see also Hille [13] and Yosida [7]).

5.4 PROBABILISTIC MECHANICS OF DISCRETE FLUIDS

It has been shown in the preceding sections that in the analysis of the dynamics of simple fluids the statistical mechanics approach is fundamental. Its main aim is to relate macroscopic quantities to the details of intermolecular effects occurring between the elements of a fluid on the microscopic level. In the present developments of fluid dynamics the formalism of molecular dynamics still uses the classical Newtonian equation of motion of an ensemble of elements (atoms, molecules), which are solved numerically and then integrated to yield the evolution of the configurational and velocity distribution. From a probabilistic mechanics point of view the molecular dynamics approach and the simulating method are both of a stochastic nature and thus the dynamics of discrete fluids can be formulated on the basis of probabilistic concepts.

(i) *Probabilistic function space in discrete fluid mechanics*

The fundamental concepts of probabilistic mechanics concerning fluids are the same as for solids. In the former the field quantities assigned to a microelement (atom, molecule) are again regarded as random variables or functions of such variables. The axiomatic definitions given in section 3.2 of Chapter 3 remain valid. Thus only those quantities pertaining to the molecular dynamics of fluids are briefly restated here. In general an element α is taken as the primitive base and designates a molecule in the discrete fluid. A smaller unit (atom), if considered, is again defined by ${}^\alpha a \stackrel{\text{df}}{=}$ an element of $\alpha \in \{\mathscr{A}\}$.

A configuration is the image of α at time t: $\mathbf{r}(\alpha, t) \in \mathbb{R}^3$ or briefly ${}^\alpha \mathbf{r} \in \mathbb{R}^3$ and a motion of α is $\alpha \stackrel{\text{df}}{=} \{{}^\alpha \mathbf{r}; -\infty < t < \infty\}$ in the discrete and time continuous case. These motions are regarded as a stochastic process $\{x(t)\}$ for each α. A mesodomain M is defined as a countable set of microelements or molecules, i.e. $M = \{{}^\alpha A\}$. Each member of the set of mathematical manifolds $\{M\} = \mathscr{M}$ represents a mesodomain. $\mathscr{M}$ is the macrodomain of the fluid body. It is of

5.4 Probabilistic mechanics of discrete fluids

particular significance to recall the definitions of volumes and densities. Thus the volume of $\alpha \in \{\mathcal{A}\}$ at time t is defined by:

$$^{\alpha}v \equiv v(\alpha, t) = \int_{\mathbf{r}(\alpha, t)} d^3\mathbf{r}; \quad ^{\alpha}v_0 \equiv v(\alpha, 0) = \int_{\mathbf{R}(\alpha, 0)} d^3\mathbf{R} \tag{5.98}$$

$\mathbf{r}$, $\mathbf{R}$ being the current and initial position vectors in the Euclidean frame, respectively. An intersecting system of molecules $\alpha, \beta, \ldots, \in \{\mathcal{A}\}$ has the volume $v(\alpha \cap \beta) = 0$. Evidently, this definition excludes the more complex case of strong interactions such as bonding to occur. By giving the elements of the structure a finite volume, it follows that the definition of mass density here is different to that given previously for the particle density at an arbitrary point $\mathbf{r}$ of the medium (relation (5.29)). Thus the mass density is defined by $^{\alpha}\rho = ^{\alpha}\mu v$ where $^{\alpha}\mu$ is a scalar element of the set M representing the mass of a molecule $\alpha \in \{\mathcal{A}\}$. For a specific mesodomain $M \in \mathcal{M}$, $^M\rho$ is defined by:

$$^M\rho = \int_{R_\rho} {^\alpha}\rho \, d\mathcal{P}(^{\alpha}\rho); \quad \int_{R_\rho} d\mathcal{P}(\rho) = 1 \tag{5.99}$$

where the Lebesgue integrand extends over the subspace $R_\rho \subset \mathcal{X}$ on the set M that is embedded in the general probabilistic function space $\mathcal{X}$ and $\mathcal{P}(^{\alpha}\rho)$ is the Lebesgue-Stieltjes measure on R_ρ. Moreover, considering the domain M on R_ρ as an open sphere of radius D or $\{^M\mathbf{r}: |\mathbf{R} - ^M\mathbf{r}| < D\}$ with respect to the Euclidean frame, where $^M\mathbf{r}$ is the position vector to the centre of mass of the meso-domain. Then the volume of this domain in accordance with Definition 9 of section 3.2 (Chapter 3) will be:

$$^Mv \sim v(M, t) \stackrel{\text{df}}{=} \int_{\mathbf{r}(M, t)} d^3\mathbf{r} \quad \text{and} \quad ^Mv_0 \equiv v(M, 0) \stackrel{\text{df}}{=} \int_{\mathbf{R}(M, 0)} d^3\mathbf{R} \tag{5.100}$$

and the mass density $^M\rho$ for this domain as given above (5.99). Analogously to the treatment of solids by recognizing the events as the basic axiomatic elements of the probabilistic formulation, the events E are the Borel sets of the Borel σ-algebra $\mathcal{F}$ of the state space $\mathcal{X}$. If $\mathcal{P}$ is a probability measure of these events $[\mathcal{X}, \mathcal{F}, \mathcal{P}]$ will define an abstract dynamical system. With reference to section 3.3, it should be noted that in a given experiment with a given finite range of measurements $\Delta\mathbf{s}$ one should introduce the events

$$E = \{^{\alpha}\mathbf{s}: \mathbf{s} < ^{\alpha}\mathbf{s} < \mathbf{s} + \Delta\mathbf{s}\}$$

giving the experimental σ-algebra $\mathcal{F}_{\text{ex}}$, whereby the convergence of results in terms of $\mathcal{P}$ and $\mathcal{P}_{\text{ex}}$ is to be achieved.

It is assumed that the components of the state vector $^{\alpha}\mathbf{s}$ are real valued functions of the geometrical and physical properties of the fluid. To represent a configurational mesodomain of the fluid body, that conceptually corresponds to the control volume in hydrodynamics, one has to use a set of state vectors or $\{^\alpha \mathbf{s}: {}^\alpha \mathbf{s}_i, \alpha = 1, 2, \ldots, N, i = 1, 2, \ldots, r\}$ and an appropriate function space. As shown for solids, it is also convenient in the considerations of fluids to use subspaces of the general state space $\mathscr{X}$. Thus, if the fluid is modelled by hard spheres and a Lennard–Jones interaction potential one can take for instance:

$$^\alpha \mathbf{r}(t) \in \mathscr{C}, \quad \mathscr{C} \subset \mathscr{X}; \quad {}^\alpha \mathbf{v}(t) \in \mathscr{V}, \quad \mathscr{V} \subset \mathscr{X}, \tag{5.101}$$

where $\mathscr{C}$ is a configuration space and $\mathscr{V}$ a velocity space. With the event structure indicated above, one obtains again subsets $E \subset \mathscr{X}$ which include the state of an element α and that may be considered analogous to (3.5, 3.6) in Chapter 3. With this interpretation one can define a class of sets $\mathscr{F}$ that form a σ-algebra and where E of $\mathscr{F}$ are Borel sets so that $\mathscr{F}, \mathscr{X}$ defines again a measurable space $[\mathscr{X}, \mathscr{F}]$. Analogously as before, it is necessary in order to analyse the random behaviour of single elements of the fluid or their collective mode of motion to choose an appropriate measure on the sets of $\mathscr{X}$ so that the probability measure for the events E is bounded. The notion of a measure is related to the fact that a dynamical variable of the flow is here, a random variable, characterized by its probability distribution function $P\{E\}$, $0 \leqslant P\{E\} \leqslant 1$ for all $E \in \mathscr{F}$ satisfying all the properties of a regular measure.

Choosing the velocity space $\mathscr{V} \subset \mathscr{X}$, for example, and a corresponding measure $\mathscr{P}^v$ satisfying the conditions of regularity, one obtains the triple $[\mathscr{V}, \mathscr{F}^v, \mathscr{P}^v]$ in the representation of the fluid system. This then may serve as a basis for the analysis of the molecular flow of simple fluids. The velocity vector $\mathbf{v} \in \mathscr{V}$, $\mathscr{V} \subset \mathscr{X}$ is then defined as a $\mathscr{P}^v$-regular measurable function in $\mathscr{V}$. The expectation or mean value of the velocity vector is given by:

$$E\{^\alpha \mathbf{v}\} = \langle {}^\alpha \mathbf{v} \rangle = \int_{\mathscr{V} \subset \mathscr{X}} {}^\alpha \mathbf{v} \, d\mathscr{P}^v \tag{5.102}$$

in which the above integral is understood in the Lebesgue–Stieltjes sense. Similarly the standard deviation is:

$$D^v = \left\{ \int ({}^\alpha \mathbf{v} - \langle {}^\alpha \mathbf{v} \rangle)^2 \, d\mathscr{P}^v \right\}^{\frac{1}{2}} \tag{5.103}$$

which makes the velocity space $\mathscr{V}$ a Hilbert space with the inner product $({}^\alpha \mathbf{v}, {}^\beta \mathbf{v}) = E\{^\alpha \mathbf{v}, {}^\beta \mathbf{v}\}$. If the expected value $E\{\mathbf{v}\} = 0$, then the norm $||\mathbf{v}||$ is equal to the standard deviation:

$$||\mathbf{v}|| = ({}^\alpha \mathbf{v}, {}^\beta \mathbf{v})^{\frac{1}{2}} = D^v. \tag{5.104}$$

5.4 Probabilistic mechanics of discrete fluids

In general D^v may satisfy the properties of a norm or in certain cases a semi-norm only. This would give rise to a specific topology on the velocity space $\mathscr{V}$. On the other hand, one can consider the topology in the configuration space $\mathscr{C}$. In this case by introducing the distance function $^{\alpha\beta}d = |^\alpha\mathbf{r} - {}^\beta\mathbf{r}|$ between two neighbouring molecules α, β, it can be regarded as a metric of the configuration space, the topology of which is that of a Hilbert space (compare with statement in section 5.2).

By considering the velocity space as a subspace of $\mathscr{X}$ and $[\mathscr{V}, \mathscr{F}^v, \mathscr{P}^v]$ a function space at any particular time t one obtains for the entire duration of the flow of a single element α a set of these function spaces or a product space. In particular for $\mathbb{R}^1 = [0, \infty)$ and each time instant $t_r \in \mathbb{R}^1, r = 1, 2, \ldots, n$ there will be a triple $[\mathscr{V}, \mathscr{F}^v, \mathscr{P}^v]$ leading to an n-fold product space in which the velocity $\mathbf{v}(^\alpha\mathbf{r}, t)$ is a measurable function. If for convenience the n-fold product space is extended to infinity, i.e. $[\mathscr{V}_\infty, \mathscr{F}_\infty, \mathscr{P}]$, the velocity function $\mathbf{v}(\mathbf{r}, t)$ becomes a time-continuous random function. This as shown earlier can also be formulated in terms of an operator L_t (eqn. (4.35)) so that:

$$L_t: \mathscr{V} \to \mathscr{V} \quad \text{for all} \quad t_r \in \mathbb{R}^+ = [0, \infty); \quad \mathbf{v}_t(\mathbf{r}) \in \mathscr{V} \tag{5.105}$$

in which the velocity field is described in terms of a random function generated by the endomorphism L_t for all $t \in \mathbb{R}^+$. This will be further discussed in section 5.5.

(ii) *General kinematics of simple fluids and interaction effects*

It is assumed in the kinematics of a simple fluid that its behaviour is Newtonian and that it consists of a collection of indistinguishable particles (molecules $\alpha, \beta, \ldots$). These assumptions are frequently made in molecular fluid dynamics. One can regard the motion of a single particle or the collective mode of motion of an ensemble of particles $\{\alpha\} = M$ (meso-domain) as a random endomorphism represented by the function $\{\mathbf{r}(\alpha, t)\} = \mathscr{A}_t \subset \mathscr{C}$ giving the position vectors to its C.M. at various times. For the collective motion a set of position vectors at time t is identified as a one-parameter family of configurational meso-domains, i.e. $\mathscr{A}_t = \{\mathbf{r}(\alpha, t) \equiv {}^\alpha\mathbf{r}(t)\}_{\alpha=1}^N \subset \mathscr{C} \subset \mathbb{R}^3$. In this sense fluid motion means that the random endomorphism G_t is such that

$$\mathscr{A}_t = G_t \mathscr{A}_0, \text{ where } \mathscr{A}_0 \equiv \mathscr{A}_t \text{ at time } t = 0.$$

The sets $\mathscr{A}_0$, $\mathscr{A}_t$ are bounded point sets in the Euclidean space $\mathbb{R}^3$ and the corresponding domains will be occupied by the same fluid elements at time t in $\mathbb{R}^3$. The concept of a Lagrangian description has been used by G. I. Taylor [175] and the notion of a geometrical transformation G_t in continuum fluid mechanics has been introduced by R. E. Meyer [176]. The latter corresponds in the probabilistic theory to the above mentioned random endomorphism.

In this description of the fluid motion, it is assumed that $r = r(\alpha, t)$ is continuous with respect to t on the domain of definition and that an inverse of G_t exists. It contains implicitly the notion, that the bounding surface of the fluid domain consists of the same type of fluid elements and that for simplicity the elements are structureless entities. This means from a micromechanics point of view, that the effect of intramolecular forces is to maintain the geometry of the molecules only. The latter constraint, however, can be relaxed. In the conventional continuum mechanics sense, it is not necessary to consider an explicit form of $r(\alpha, t)$ for the description of the fluid motion. The latter is rather characterized by the velocity field v as a function of (r, t). All corresponding random functions of the motion are then defined for a fluid domain as a family of associated random transformations during an open time interval such that:

$$G_t\{^\alpha r(0)\} = \{^\alpha r(t)\} \qquad (5.106)$$

It is further assumed that there exists a class $C^n(\mathscr{C})$ with continuous derivatives of order $\leqslant n$ with respect to r, and t for a chosen time interval $t_1 < t < t_2$.

It is well known that the Lagrangian representation in terms of $r(\alpha, t)$ in the hydrodynamic theory is the solution of the deterministic differential equation:

$$\frac{\partial^\alpha r(t)}{\partial t} = {}^\alpha v({}^\alpha r, t) \qquad (5.107)$$

with the initial condition ${}^\alpha r(0) = r_0$ in $\mathscr{A}_0$ and the velocity vector $v \in C^\infty(G_t \mathscr{A}_0)$. Evidently the connection of this description to the Eulerian one is given by:

$$h({}^\alpha r, t) \equiv f[{}^\alpha r(\alpha, t)] \qquad (5.108)$$

and
$$Df/Dt \equiv \frac{\partial}{\partial t} h({}^\alpha r, t) = \left[\frac{\partial}{\partial t} + v \cdot \text{grad}\right] f \qquad (5.109)$$

in which Df/Dt is the material derivative of f. Since the rate of change with time of characterizing the motion of $\alpha \in M$ passing through the position of α at time t $r(\alpha, t)$ differs generally from $\partial f/\partial t$ the material derivative must be taken. Equation (5.107) in probabilistic mechanics is, however, a stochastic differential equation with the same initial conditions and for which a solution exists (see also Yosida [7], Bharucha-Reid [18]). In section 5.5 the motion of fluid elements will be analysed in terms of the Markov process theory. Moreover, the identification of a Gibbsian random field with a Markov random field permits the introduction of interaction effects between elements of the discrete fluid in a natural way. Before dealing with this approach it is, however, necessary to discuss briefly interaction effects in simple fluids.

Interactions between neighbouring elements of a structured medium in general have already been discussed in section 3.5 of Chapter 3. It has been

5.4 Probabilistic mechanics of discrete fluids 147

shown that interactions have both attractive and repulsive components. Simple fluids are conventionally modelled by hard-spheres, square-well and more often on the basis of the Lennard–Jones potential $\phi(r)$. The latter can be simply written as:

$$\phi(r) = 4\varepsilon\left[\left(\frac{r_e}{r}\right)^m - \left(\frac{r_e}{r}\right)^n\right] \quad (5.110)$$

where ε is the well-depth, σ here is the intermolecular distance at equilibrium. The exponents m and n are normally taken to be equal to 12 and 6, respectively, r is the intermolecular distance. Accordingly, the hard-sphere potential is given by:

$$\phi(r) = \begin{Bmatrix} \infty, 0 < r < r_e \\ 0, r > r_e \end{Bmatrix}. \quad (5.111)$$

As mentioned earlier, direct structural information concerning the molecular arrangement in liquids is obtained from X-ray and fast neutron-scattering experiments. Theoretical results of the molecular-dynamics analysis are frequently checked by computer simulations of which the previously mentioned Monte-Carlo method is the most often used. In the adopted notation, the interaction potential between molecules α, β designated by $^{(\alpha\beta)}\phi$ is a function of the intermolecular distance $^{\alpha\beta}d = |^\alpha\mathbf{r} - {}^\beta\mathbf{r}|$. It is evident, that it will be accurate only for short-range interactions. Thus by assuming a hard sphere model for the idealized fluid, it can be regarded as an intrinsic or reference potential. For real fluids, however, where bond association with other molecules has to be admitted within a certain neighbourhood of the molecule, one can write an overall potential in the form of:

$$^{(\alpha\beta)}\phi(\mathbf{d}) = {}^{(\alpha\beta)}\phi_0 + {}^\beta\phi \quad (5.112)$$

in which $^{(\alpha\beta)}\phi_0$ is the intrinsic potential and $^\beta\phi$ is the part that accounts for bonding with surrounding molecules. This part of the interaction potential $^{(\alpha\beta)}\phi$ will be non-zero only for a small range of relative positions and orientations of the molecules (Fig. 1, Chapter 3). Since the reference potential $^{(\alpha\beta)}\phi_0$ contains a strongly repulsive part that prevents two molecules from coming too close to one another, the distance $^{\alpha\beta}d$ corresponds to that of a short-range order or a minimum distance. In the present theory it is seen as a distance function $d(\alpha, \beta) \equiv d(^\alpha\mathbf{r}, {}^\beta\mathbf{r})$ or as a metric on the corresponding topological configuration space $\mathscr{C} \subset K$. It will be further discussed subsequently.

In this context one can distinguish various density regimes in real fluids. By using the number density $n(\mathbf{r}, t)$ introduced at the beginning of this chapter and a characteristic length d_c for the range of the interaction potential $^{(\alpha\beta)}\phi$, it may be stated that for the low-density regime the condition $n(\mathbf{r}, t)d_c^3 \ll 1$ holds. In the case of dense gases or at liquid densities this condition becomes approximately $n(\mathbf{r}, t)d_c^3 \gtrsim 0.1$ (see, for instance, Batchelor [178]). Since in the probabilistic mechanics the concept of configurational mesodomains is employed, which for an idealized fluid is synonymous to the control volume of hydrodynamics, a characteristic length L of such a fluid domain is significant. It must be such that $L \gg l_d$ or the diffusion length associated with the diffusion of a microelement out of the particular mesodomain [181]. Correspondingly various time scales of the fluid flow can be considered: a characteristic macrotime $\tau_\mu \sim L^2/6D$ (where D is the conventional diffusion constant) and a characteristic microtime at which the particle velocity distribution occurs. Evidently the macrotime scale must be much larger than the microtime scale (compare with the discussion on microdynamics section 4.7, Chapter 4).

5.5 MARKOV THEORY IN THE MECHANICS OF DISCRETE FLUIDS

(i) *Markov process approach*

The probabilistic mechanics of discrete fluids was briefly discussed in section 5.4. Thus the evolution of the velocity field $\mathbf{v} = \mathbf{v}(\mathbf{r}, t)$ for instance can be represented in terms of the triple $[\mathscr{V}, \mathscr{F}^v, \mathscr{P}^v]$ with an associated operator $T_t: \mathscr{V} \to \mathscr{V}$ parametrized by $t \in \mathbb{R}^1 = [0, \infty]$ so that $\mathbf{v}_t(\mathbf{r}) \in \mathscr{V}$ is a stochastic process. Completely analogous to the representation of the behaviour of solids by means of stochastic processes, the latter can also be employed in the description of the molecular dynamics of fluids [179]. Here again the Markov theory provides possible models for the probabilistic approach. Thus consider the state space $S \equiv \mathscr{X}$ (Def. 11 of Chapter 3) to consist of points S or x representing possible observations of configurational or other dynamical quantities at any given fixed time. The events ^sE for which a probability is well defined are the elements of a Borel field $\mathscr{F}$ of subsets of $\mathscr{X}$. The stable generating mechanism of a Markov process characterized by its transition function $P(t, x, E)$ assumed to be $\mathscr{F}$-measurable is then a function of x for each set or event ^sE in $\mathscr{F}$ and a probability measure on the Borel field $\mathscr{F}$ for each x in $\mathscr{X}$. Thus $P(t, x, E)$ represents the probability, that an outcome at time t_{r+1} of the Markov process will fall in the event set ^sE given that at time t_r the observation x was made. Thus considering the velocity field $\mathbf{v}(\mathbf{r}, t)$ and using conditional probabilities there will be for each transformation T_t a corresponding probability measure $\mathscr{P}^v\{E_{r+1}|E_r\}$ such that, whenever

one has
$$\left.\begin{aligned}T_t E_r &= E_{r+1} \\ \mathscr{P}^v\{E_{r+1}\} &= \mathscr{P}^v\{E_{r+1}|E_r\}\mathscr{P}^v\{E_r\}.\end{aligned}\right\} \quad (5.113)$$

Similarly as before this can be generalized to:
$$\mathscr{P}^v\{E_n\} = \mathscr{P}^v\{E_i\} \prod_{r=1}^{n-1} \mathscr{P}^v\{E_{r+1}|E_r\} \quad (5.114)$$

5.5 Markov theory in the mechanics of discrete fluids

which is valid for any sequence $E_1, E_2, \ldots E_r, E_n$ corresponding to the time sequence $t_1 < t_2 < \ldots < t_r < t_n$ and a set of conditional probabilities $\mathscr{P}\{E_{r+1}|E_r\}, r = 1, 2, \ldots, n-1$. This can be written more explicitly as:

$$\mathscr{P}^v\{E_{r+1}, t_{r+1}\} = \mathscr{P}^v\{t_{r+1}, t_r\}\mathscr{P}^v\{E_r, t_r\} \qquad (5.115)$$

where E_r corresponds to $t_r \in \mathbb{R}^1$ and E_{r+1} to $t_{r+1} > t_r$.

Analogous to the case of solids [180], the time evolution of the velocity probability distributions in terms of a matrix differential equation can be given as:

$$\underline{P}^v\{t+s\} = \underline{P}^v\{t\}\underline{P}^v\{s\} \qquad (5.116)$$

which holds for any small time interval $[t, s]$ during the uniform flow of the discrete fluid. This relation indicates again the semi-group property of the system. As before, it can also be represented by a one-parameter semi-group of the linear transformation $T_t: t \geq 0, t \in \mathbb{R}^1$ given earlier. These operators are defined in the function space $C(\mathscr{V}) \subset \mathscr{X}$ by:

$$T_t[f(w)] = \int_{\mathscr{V} \subset \mathscr{X}} f(u) P(t, w, du), \quad f \in C(\mathscr{V}). \qquad (5.117)$$

T_t will satisfy the following conditions:

$$T_t T_s = T_{t+s} \quad \text{for all} \quad t, s > 0 \text{ in } \mathscr{V} \subset \mathscr{X}. \qquad (5.118)$$

The linear semi-group $\{T_t; t \geq 0\}$ is contracting, if $\|T_t f\| \leq \|f\|, f \in C(\mathscr{V})$. For considerations of the Markov process of a single fluid element $\alpha \in M$, the weak-limit is important. This as mentioned in Chapter 2 is given by:

$$w\text{-}\lim_{t \downarrow t_0} T_t[f(w)] = f(w) \quad \text{for every} \quad f \in C(\mathscr{V}) \qquad (5.119)$$

since bounded observables or events induce a weak topology on $\mathscr{V}$ only and which is closely related to Markov processes.

It has to be recognized that generally an outcome in the velocity space $\mathscr{V}$ is a vector-valued quantity. Thus it becomes necessary to consider a multi-dimensional random process for the description of the flow of a single particle or microelement of the fluid. However, for the condition of a uniform flow this can be simplified by considering a scalar process, i.e. where the velocities are characterized by $|v|$ only.

Thus returning to relation (5.116) the transition matrix written as $P_{ij}(t)$ represents the transition of a fluid element α from a state i at time t_r to a state j at time t_{r+1}. As shown earlier in dealing with deformations of solids in adopting a time-homogeneous Markov process the following limits exist:

$$\lim_{t \to 0^+} \frac{1 - P_{ii}(t)}{t} = -\lambda_{ii} < \infty; \quad \lim_{t \to 0^+} \frac{P_{ij}(t)}{t} = \lambda_{ij}$$

in which the λ_{ij} are the elements of the transition intensity matrix Q_{ij} that has also been denoted by $\underline{Q}$. Hence the evolution of properties in the velocity space $\mathscr{V} \subset \mathscr{X}$ can be expressed by the Kolmogorov matrix differential equation:

$$\frac{d\underline{P}^v(t)}{dt} = \underline{Q}^v \underline{P}^v(t). \tag{5.120}$$

During uniform flow, i.e. excluding turbulence, the Markov theory applies, but the pertinent process will be a Poisson-type process. This can be readily recognized by considering the uniform flow corresponding to independent events in the terminology of probability theory, in which such a process has the property that a number of events occurring within the time t is a Markov process with the invariant measure $\mathscr{P}^v$. The corresponding transition intensities are then given by:

$$\lambda_{ij} = \begin{cases} \lambda & \text{for } j = i+1 \\ 0 & \text{for } j \neq i+1 \end{cases} \quad (i = 0, 1, 2, \ldots). \tag{5.121}$$

Thus only direct transitions from a state $i \to j = i+1$ $(i = 0, 1, 2, \ldots)$ can take place. It is possible to consider also n-step transition probabilities based on recurrence relations (see [19, 28]), which are required in simulation techniques briefly indicated below. It can be shown that from the Kolmogorov differential equation involving the functions $P_{ij}(t)$ in (5.120) a system of differential equations can be formed, the solution of which is given by:

$$\left. \begin{array}{l} P_{ij}(t) = \dfrac{(\lambda t)^{j-i}}{(j-i)!} e^{-\lambda t} \quad \text{for } j \geq i \\[2pt] \phantom{P_{ij}(t)} = 0 \quad \text{for } j < i \end{array} \right\} \tag{5.122}$$

representing a Poisson process. It is evident that the notion of independent events introduced above must also be considered for an ensemble of interacting fluid elements within a given mesodomain. This necessitates the use of a common transition function to be introduced in section (iii) below.

(ii) *Remarks on computer simulation*

As already mentioned in Chapter 4 computer simulations become essential, when dealing with complex systems. They replace on the one hand actual experiments on such systems and on the other permit to obtain quantitative results, which cannot be obtained by analytical methods. So far as discrete fluids are concerned, these techniques are primarily used for the determination of physical properties. However, in order to gain insight into the dynamical behaviour of fluid systems an approach based on Markov principles has to be developed. These techniques are therefore based on probabilistic concepts in which the randomness of the physical structure on the microscale is accounted for, implicitly. This is in contrast with the philosophy of the deterministic

5.5 Markov theory in the mechanics of discrete fluids

molecular dynamics approach, that is entirely based on the Newtonian equations of motions (see, for instance, Bellemans [162]).

Thus, in the well-known Monte Carlo method as applied to stationary processes a system of N-interacting elements with a specified interaction potential is given a set of arbitrarily chosen initial coordinates and a sequence of configurations of the elements is generated by admitting random displacements to occur. However, a decision is required whether a particular configuration can be accepted or must be rejected, since the results must comply to configurations that correspond to an unknown probability density function of the ensemble under consideration, i.e. canonical, etc. The particular method in the Monte Carlo simulation used in liquid-state physics goes back to Metropolis *et al.* [161]. It consists of generating a set of molecular configurations for a specific state i corresponding to the random displacements of an N-particle ensemble. If the configurations are discrete and finite in number, the theoretical form for the canonical ensemble average of a dynamical variable or function f can be approximated by:

$$\langle f \rangle_{(i)} = \frac{\sum_{i=1}^{m} \exp[-\beta U_N(i)] f(i)}{\sum_{i=1}^{m} \exp[-\beta U_N(i)]} \quad (5.123)$$

in which $U_N(i)$ is the potential energy of the configuration in the state i symbolically written as (i), m the total number of configurations in the set and β the Boltzmann factor. Since m is assumed to be large, it is necessary for computational purposes to use a smaller number of configurations for the state (i) or $r \ll m$. By using the probability distribution function $P(j)$ of the form:

$$P(j) = \frac{\exp[-\beta U_N(j)]}{\sum_{i=1}^{r} \exp[-\beta U_N(i)]} \quad (5.124)$$

a selection of possible configurations in the set $\mathscr{A}_t \subset \mathscr{C}$ can be made and the ensemble average of the function f can be taken in an approximate form as:

$$\langle f \rangle \cong \frac{1}{r} \sum_{i=1}^{r} f(i). \quad (5.125)$$

From the point of view of Markov theory, it is necessary for the above approximation to generate a Markov chain so that the unweighted average of f over all states converges for a sufficiently large number, $r \ll m$ to the theoretical value of f for a canonical ensemble. This may be achieved by using a recurrent chain. Recalling the Chapman–Kolmogorov equation for the

152   Probabilistic Mechanics of Fluids

transition probabilities given earlier, i.e.:

$$P_{ij}(t_r+t_s) = \sum_k P_{ik}(t_r) P_{kj}(t_s)$$

which corresponds to a one-step transition from a state i to j, one can also use an n-step transition probability. Thus, if the homogeneous process is considered only at discrete time instants, as required in the Monte-Carlo method, then $t = n \cdot h (n = 0, 1, 2, \ldots; h > 0)$ and the probabilities $P_{ij}(nh)$ for n-steps are uniquely defined from the one-step probabilities ($P_{ij} = P_{ij}(h)$) such that:

$$\left. \begin{array}{l} P_{ij}(nh) = \sum_k P_{ik} P_{kj}[(n-1)h] = \sum_k P_{ik}[(n-1)h] P_{kj} \\ (i,j = 1, 2, \ldots) \text{ for all } n = 1, 2, \ldots \end{array} \right\} \quad (5.126)$$

Moreover, the transition probabilities of a homogeneous chain satisfy the continuity property, whereby:

$$P_{ij}(0) = \lim_{h \to 0} P_{ij}(h) = \begin{cases} 1 & \text{for } j = i, \\ 0 & \text{for } j \neq i. \end{cases} \quad (5.127)$$

These quantities will also be continuously differentiable and the following limits will exist:

$$\lim_{h \to 0} \frac{P_{ij}(h) - P_{ij}(0)}{h} = q_{ij}; \quad (i,j = 1, 2, \ldots) \quad (5.128)$$

where q_{ij} for $i \neq j$ is finite. In the present case, by assuming that the transition matrix $\underline{P}$ is independent of time and that its elements satisfy the conditions:

$$P_{ij} \geq 0; \quad \sum_{j=1}^{r} P_{ij} = 1$$

a recurrence relation for the n-step transition probabilities denoted by $P_{ij}^{(n)}$ is then as follows:

$$P_{ij}^{(n)} = \sum_{k=1}^{r} P_{ik}^{(n-1)} P_{kj}. \quad (5.129)$$

If all states in the above forms (5.128) and (5.129) belong to the same ergodic class of sets (see also Yosida and Kakutani [125]), then for all pairs i, j a value for n in (5.128) can be found such that $P_{ij}^{(n)} > 0$ with the limit:

$$P_j^* = \lim_{n \to \infty} P_{ij}^{(n)} \quad \text{for all } j \quad (5.130)$$

that are independent of i. Hence the following conditions will be satisfied:

$$P_j^* > 0; \quad \sum_{j=1}^{r} P_j^* = 1; \quad P_j^* = \sum_{i=1}^{r} P_i^* P_{ij}. \quad (5.131)$$

Since in the simulation method the limit is prescribed (5.125), it follows from relations (5.129–5.131) that:

$$P_j^* \equiv P(j) \equiv \text{eqn. (5.124)}.$$

It is apparent that the stochastic matrix, which has to be established for performing the simulation method, must be chosen so as to satisfy the condition given by (5.131).

A more detailed discussion on the computer-simulation procedure concerning molecular fluids is given amongst others by Hansen [159] and McDonald [155]. The accuracy that can be achieved by this technique and in particular the effect of truncating the employed interaction potentials is also discussed by Croxton [182].

(iii) Interacting Markov processes; Markov fields

It has been indicated in paragraph (i) of this section that the Markov approach in the description of the motion of a single particle or element of the fluid leads to a Poisson-type process (eqn. (5.122)). Due to the interaction effects between neighbouring elements α, β it is to be expected that the formulation should be based on interacting Markov processes. Thus, in the simplest way from the standpoint of classical lattice statistical mechanics, one can introduce a common transition probability function for two neighbouring configurations i, j of the particles of the fluid. Such an approach has been considered by Preston [183] and in terms of coupled Markov chains by Spitzer [184] and Holley [185].

Thus, in the adopted notation of the present study and considering the configuration space $\mathscr{C} \subset \mathscr{X}$ a common transition function denoted by $P_t(i,j)$ can be introduced, which is valid on a configurational mesodomain $\mathscr{A}_t$ of the fluid body. For a countably finite set of all possible configurations on that domain with independent Markov processes on it, the function $P_t(i,j)$ will be such that:

$$\sum_{i \in \mathscr{A}_t} P_t(i,j) = \sum_{j \in \mathscr{A}_t} P_t(i,j) = 1, i, j \in \mathscr{C}, t \geqslant 0 \quad (5.132)$$

where $P_t(i,j)$ is assumed to be an invariant measure on $\mathscr{C}$. Hence for the uniform flow of a discrete fluid considering the description of the motion for a single element (eqn. (5.122)) the probability distribution for elements $\alpha = 1, 2, \ldots, N$ and a transition intensity $\lambda > 0$ this distribution can be taken as:

$$\mathscr{P}^m(B) = e^{-m\lambda} \prod_{i=1}^{m} \frac{\lambda^{N_i}}{N_i!}; \; N_i \in \mathscr{A}_t, \mathscr{A}_t \subset \mathscr{C}, i = 1 \ldots m \quad (5.133)$$

in which $B \in \mathscr{F}^{\mathscr{C}}$ (σ-algebra in the configuration space $\mathscr{C}$) and $\{N_i\}$ is defined by the Borel set B. $\mathscr{P}^m$ is an invariant measure for the microelements $\alpha = 1, \ldots, N$ undergoing a transition in one-step from state $i = 1, 2, \ldots, m$. Thus the uniform flow of a simple liquid can be described in terms of the Markov theory and the invariant probability measure $\mathscr{P}^m$ in the corresponding topological space. This measure is seen to depend largely on the interaction effect between elements of the fluid.

It is of greater interest in the present analysis to establish a formulation for the flow of discrete fluids in terms of a known or assumed pair-potential. This can be achieved on the basis of the more recent advances in Markov theory. In accordance with Postulate 1 (Chapter 3) of the present theory, that an ensemble of microelements (molecules) is regarded as a Gibbsian ensemble, the notion of a Gibbsian random field is important. For the definition of such a random field one needs an interaction potential $\phi(^\alpha\mathbf{r}, ^\beta\mathbf{r})$: $\mathscr{C} \times \mathscr{C} \to \mathbb{R}^1$ satisfying the following properties:

$$\left.\begin{array}{l} \text{(i) symmetry:} \quad \phi(^\alpha\mathbf{r}, ^\beta\mathbf{r}) = \phi(^\beta\mathbf{r}, ^\alpha\mathbf{r}), \\ \text{(ii) homogeneity:} \quad \phi(^\alpha\mathbf{r}, ^\beta\mathbf{r}) = \phi(0, ^\beta\mathbf{r} - ^\alpha\mathbf{r}), \\ \text{(iii) Nearest-neighbour property:} \\ \qquad \phi(^\alpha\mathbf{r}, ^\beta\mathbf{r}) = 0, \text{ if } |^\beta\mathbf{r} - ^\alpha\mathbf{r}| > \delta \end{array}\right\} \quad (5.134)$$

where δ in condition (iii) denotes the range of interactions as discussed previously in section 3.5 of Chapter 3. Thus the triplet $[\mathscr{C}, \mathscr{F}, \mathscr{P}]$ for $M = \{\alpha\}$ on $D \subset Z^3$, where D corresponds to the conventional concept of a control volume and together with the above defined potential $\phi(^\alpha\mathbf{r}, ^\beta\mathbf{r})$ designates a Gibbsian random field. For a more specific definition of such a random field, a boundary value function f on ∂D is, however, required (see also Spitzer [184] [177]). This function represents an arbitrary map $f: \partial D \to \{0, 1\}$. For convenience one can extend this description to a map $\bar{f}: \bar{D} \to \{0, 1\}$, $\bar{D} = D \cup \partial D$ so that:

$$\bar{f}(^\alpha\mathbf{r}) = \begin{cases} f(^\alpha\mathbf{r}) & \text{for } ^\alpha\mathbf{r} \in D \\ \partial f(^\alpha\mathbf{r}) & \text{for } ^\alpha\mathbf{r} \in \partial D \end{cases} \quad (5.135)$$

where $^\alpha\mathbf{r}$, $^\beta\mathbf{r}$, ... designate any lattice sites at which the particle α is positioned. For a given control volume D and an interaction potential satisfying conditions (5.134) and a boundary-value function defined in (5.135) the Gibbsian random field will be specified with:

$$P(^\alpha\mathbf{r}) = Z^{-1} \exp\left[-\frac{1}{2} \sum_{^\alpha\mathbf{r} \in \bar{D}} \sum_{^\beta\mathbf{r} \in \bar{D}} \bar{f}(^\alpha\mathbf{r}) \bar{f}(^\beta\mathbf{r}) \phi(^\alpha\mathbf{r}, ^\beta\mathbf{r})\right] \quad (5.136)$$

in which $Z = Z(D)$ is the equivalent of the partition function of a canonical

5.5 Markov theory in the mechanics of discrete fluids

ensemble. It is important to note, that the findings of several researchers on the equivalence of the above Gibbsian random field and a Markov random field leads to a representation of the flow of discrete fluids in terms of the Markov theory. In particular a theorem by Spitzer on this equivalence is of special interest here, which can be stated as follows:

"Every Markov random field on a given domain M with a boundary ∂M or a boundary value function f is a Gibbsian random field with the boundary ∂M (or a boundary value function f) and vice versa."

It is of interest to note, that the potential defined by its properties (5.134) excludes the more distant than the nearest-neighbour interactions. For particle systems in which such non-local effects may be significant, a formulation is still possible but would involve the d-Markov fields as discussed in Chapter 2. The probabilistic analysis of simple fluids presented here can also be employed in the description of molecular fluids. In conclusion, it may be stated that the probabilistic description of the behaviour of discrete media can be given in an analytically unified manner analogous to the classical deterministic formulation in mechanics. However, for the use of probabilistic concepts and principles the modern tools of functional analysis have to be employed. In this respect, the basic notion of material and evolution operators for discrete materials are at the foundations of probabilistic mechanics. Fundamentally, a complete description of the behaviour of discrete media would also require the formulation of an appropriate random-field theory.

Bibliography

1. A. N. KOLMOGOROV, *Grundbegriffe der Wahrscheinlichkeitsrechnung*, Springer, Berlin (1933).
2. A. N. KOLMOGOROV, *Foundations of the Theory of Probability*, Chelsea, New York (1950).
3. N. BOURBAKI, *Topologie Générale*, Herman, Paris (1958).
4. K. KURATOWSKI, *Topology I & II*, Academic Press–PWN, New York–Warszawa (1966–1968).
5. A. N. KOLMOGOROV and S. V. FOMIN, *Elements of the Theory of Functions and Functional Analysis*, Vol. I, Graylock, Rochester and Albany (1957).
6. K. KURATOWSKI and A. MOSTOWSKI, *Set Theory*, North Holland, Amsterdam (1976).
7. K. YOSIDA, *Functional Analysis*, Springer, Berlin (1978).
8. J. L. KELLEY, *General Topology*, van Nostrand, New York (1955).
9. Y. CHOQUET-BRUHAT and C. DE WITT-MORETTE and M. DILLARD-BLEICK, *Analysis, Manifolds and Physics*, North-Holland, Amsterdam (1977).
10. A. N. KOLMOGOROV and S. V. FOMIN, *Elements of the Theory of Functions and Functional Analysis*, Vol. II, Graylock, Rochester and Albany (1961).
11. P. R. HALMOS, *Measure Theory*, van Nostrand, New York (1950).
12. K. R. PARTHASARATHY, *Probability Measures on Metric Spaces*, Academic Press, New York (1967).
13. E. HILLE, *Functional Analysis and Semi-Groups*, Publ. Amer. Math. Soc. (1948).
14. R. VON MISES, *Wahrscheinlichkeitsrechnung*, Deuticke, Leipzig (1931).
15. B. C. VAN FRASSEN, "Foundations of probability: a modal frequency interpretation", in *Problems in the Foundations of Physics* (ed. G. T. DI FRANCIA), North-Holland, Amsterdam (1979).
16. S. P. GUDDER, *Stochastic Methods in Quantum Mechanics*, North-Holland, New York (1979).
17. A. RÉNYI, *Foundations of Probability*, Holden-Day, San Francisco (1970).
18. A. T. BHARUCHA-REID, *Random Integral Equations*, Academic Press, New York (1972).
19. YU. V. PROHOROV and YU. A. ROZANOV, *Probability Theory*, Springer, New York (1969).
20. A. A. BOROWKOW, *Wahrscheinlichkeitstheorie*, Birkhäuser Verlag, Basel (1976).
21. J. NEVEU, *Mathematical Foundations of the Calculus of Probability*, Holden-Day, San Francisco (1965).
22. M. LOÈVE, *Probability Theory*, Van Nostrand, New York (1962).
23. V. S. PUGACHEV, *Theory of Random Functions*, Pergamon Press, Oxford (1965).
24. M. ROSENBLATT, *Markov Processes. Structure and Asymptotic Behavior*, Springer, New York (1971).
25. B. V. GNEDENKO, *Theory of Probability*, Chelsea, New York (1963).
26. J. C. Samuels, "Elements of stochastic processes", in *Continuum Physics* (ed. by A. C. ERINGEN), Academic Press, New York (1971).
27. M. FRÉCHET, *Generalités sur les Probabilités*, Gauthier-Villars, Paris (1950).
28. W. FELLER, *An Introduction to Probability Theory and its Applications*, Vols. I and II, Wiley, New York (1966).
29. A. KHINTCHINE, *Asymptotic Laws in Probability Theory*, ONTI (1935).
30. A. M. YAGLOM, *Stationary Random Functions*, Prentice-Hall, Englewood Cliffs (1962).
31. E. PARZEN, *Modern Probability Theory and its Application*, Wiley, New York (1960).
32. J. MIKUSIŃSKI, *The Bochner Integral*, Birkaüser Verlag, Basel (1978).
33. J. L. DOOB, *Stochastic Processes*, Wiley, New York (1961).
34. N. WIENER, *Non-Linear Problems in Random Theory*, The Technology Press of the Massachusetts Institute of Technology and New York; Wiley (1958).

35. D. R. AXELRAD, *Micromechanics of Structured Materials*—Sectional Invited Lecture, 14th Congress of Theoretical and Applied Mechanics, IUTAM Delft, The Netherlands (1976).
36. D. R. AXELRAD, *Micromechanics of Solids*, Elsevier-PWN, Amsterdam–Warszawa (1978).
37. I. M. GEL'FAND and N. YA. VILENKIN, *Generalized Functions*, Vols. I–IV, Academic Press, New York (1964).
38. YU. A. ROZANOV, *Stationary Random Processes*, Moscow, Fizmatgiz (1963) (Russian).
39. K. ITO and H. MCKEAN, *Diffusion Processes and their Sample Paths*, Springer–Heidelberg–New York (1965).
40. L. SCHWARTZ, *Theorie de Distributions*, Hermann, Paris (1950).
41. A. N. KOLMOGOROV, "The local structure of turbulence in incompressible viscous fluid for very large Reynolds numbers", *Dokl. Akad. Nauk SSSR* **30**, 301–305 (1941), **31**, 538–540 (1941).
42. A. A. MARKOV, *Calculus of Probability*, 4th ed., Moscow (1924), (Russian).
43. G. K. BATCHELOR, *The Theory of Homogeneous Turbulence*, Cambridge University Press, London–New York (1953).
44. W. GUZ, "Markovian processes in classical and quantum statistical mechanics", *Rep. Math. Phys.* **7** No. 2, 205–214 (1975).
45. E. DYNKIN, *Markov Processes*, Vols. I and II, Academic Press, New York (1965).
46. R. M. BLUMENTHAL and R. K. GETOOR, *Markov Processes and Potential Theory*, Academic Press, New York (1968).
47. D. REVUZ, *Markov Chains*, North-Holland, Amsterdam (1975).
48. G. D. BIRKHOFF, "Proof of the ergodic theorem", *Proc. N.A.S.* **17**, 656–660 (1931).
49. J. VON NEUMANN, "Proof of the quasi-ergodic hypothesis", *Proc. N.A.S.* **18**, 70 (1932).
50. A. KOSSAKOWSKI, "Time evolution in isolated and non-isolated quantum mechanical systems", *Rep. Math. Phys.* **3**, 247 (1973).
51. G. MACKEY, *The Mathematical Foundation of Quantum Mechanics*, Benjamin, New York (1963).
52. P. L. BUTZER and H. BERENS, *Semi-Groups of Operators and Approximation*, Springer, New York (1967).
53. R. L. DOBRUSHIN, "Description of a random field by means of conditional probabilities and the conditions governing its regularity", *Theory of Prob. and its Appl.* **10**, 193–213 (1969).
54. M. B. AVERINTSEV, "On a method of describing discrete parameter random fields", *Problemy Peredachi Informatsii* **6**, 100–109 (1970).
55. YU. A. ROZANOV, "On Gaussian fields with given conditional distributions", *Theory of Prob. and its Appl.* **12**, 381–391 (1967).
56. P. LÉVY, "A special problem of Brownian motion and a general theory of Gaussian random function", *Proc. 3rd Berkeley Symp. Math. Stat. and Prob.* **2**, 133–175 (1956).
57. E. WONG, *Stochastic Processes in Information and Dynamical Systems*, McGraw-Hill, New York (1971).
58. M. J. BERAN, *Statistical Continuum Theories*, Interscience Publ., New York (1968).
59. I. A. KUNIN, *Theory of Solid Media with Microstructure*, Nauka, Moscow (1975) (Russian).
60. D. R. AXELRAD, "The Mechanics of Discrete Media", General Lecture, Continuum Models of Discrete Systems, Symp. Freudenstadt (1979).
61. A. I. KHINTCHINE, *Mathematical Foundations of Statistical Mechanics*, Dover, New York (1949).
62. J. KAMPÉ DE FÉRIET, "Statistical mechanics of discrete media", *Proc. Symp. Appl. Math.*, pp. 165–198, Amer. Math. Soc., Providence, R. I. (1962).
63. M. KAC, "Probability and related topics in physical sciences", *Lectures in Appl. Math.*, Vol. I, Interscience, New York (1959).
64. H. MARGENAU and N. R. KESTNER, *Theory of Intermolecular Forces*, Pergamon Press, Oxford (1969).
65. J. YVON, *Correlations and Entropy in Classical Statistical Mechanics*, Pergamon Press, Oxford (1969).
66. J. O. HIRSCHFELDER, C. F. CURTISS and R. B. BIRD, *Molecular Theory of Gases and Liquids*, John Wiley, New York (1954).
67. A. D. BUCKINGHAM, in *Intermolecular Interactions: From Diatomics to Bipolymers* (ed. B. PULLMAN), Wiley, Chichester (1977).

68. A. D. BUCKINGHAM, in *Microscopic Structure and Dynamics of Liquids* (ed. J. DUPUY and A. J. DIANOUX), Plenum Press, New York (1977).
69. P. SCHUSTER, G. ZUNDEL and C. SANDORFY (Ed.), *The Hydrogen Bond*, Vols. I, II and III, North-Holland, Amsterdam (1976).
70. M. BORN and Y. HUANG, *Dynamical Theory of Crystal Lattices*, Oxford (1954).
71. A. MOTT and B. JONES, *Properties of Metals and Alloys*, Oxford (1936).
72. J. P. HANSEN and J. R. MCDONALD, *Theory of Simple Liquids*, Academic Press, London (1976).
73. H. MATSUDA and Y. HIWATARI, "The effect of interatomic potential on the feature of solid–liquid phase transition", in *Cooperative Phenomena* (ed. H. HAKEN and M. WAGNER), Springer, New York (1973).
74. A. T. BHARUCHA-REID, *Random Integral Equations*, Academic Press, New York (1972).
75. K. ITO, "On stochastic differential equations", *Mem. Amer. Math. Soc.*, No. 4 (1951).
76. L. ARNOLD, *Stochastiche Differentialgleichungen-Theorie und Anwendung*, Oldenburg, München 74.
77. O. HANŠ, "Random operator equations", *Proc. 4th Berkeley Symp. on Math Stat. and Prob.*, Vol. II, pp. 185–202 (1960).
78. A. ŠPAČEK, "Zufällige Gleichungen", *Czechoslovak Math. J.* 5, 462–466 (1955).
79. H. P. MCKEAN, Jr., *Stochastic Integrals*, Academic Press, New York (1969).
80. M. FRÉCHET, "Les elements aleatoires de nature quelconque dans un espace distancié", *Ann. Inst. H. Poincaré*, 10, 215–310.
81. E. MOURIER, "L-random elements and L*-random elements in Banach spaces", *Proc. 3rd Berkeley Symp. on Math. Stat. and Prob.*, pp. 231–242 (1955).
82. O. HANŠ, "Generalized random variables", *Trans. 1st Praque Conf. on Information Theory, Stat. Decision Functions and Random Processes*, pp. 61–103 (1956).
83. D. R. AXELRAD and J. W. PROVAN, *Arch. Mech. Stos.* 25, 811 (1973).
84. D. R. AXELRAD, "Rheology of structured media", *Arch. Mech. Stos.* 23, 31 (1971).
85. D. R. AXELRAD, *Random Theory of Deformation of Structured Media*, Lecture 7th Int. Cent. Mech. Sci., Udine, Italy (1971).
86. Y. M. HADDAD, "Response Behavior of a Two-dimensional Fibrous Network", Ph.D. Thesis, McGill University, Montreal, Canada (1975).
87. J. J. MOREAU, "Sur l'evolution d'un system elasto-viscoplastique", *C. R. Acad. Sci.*, Ser. A, 272, 118 (1971).
88. D. R. AXELRAD, J. W. PROVAN and S. BASU, "Analysis of semi-group property and the constitutive relations of structured solids", in *Symmetry, Similarity and Group Theoretic Methods in Mechanics*, Proc. of Symposium, Calgary, Canada (1974).
89. D. R. AXELRAD, "Micromechanics of discrete systems", *Arch. Mech. Stos.* 28, 299 (1973).
90. D. R. AXELRAD, "Rheology of discrete media", in *7th Int. Congress on Rheology, Proc.* (ed. C. KLASON and J. KUBAT), Gothenburg, Sweden (1976).
91. RI. PERALTA-FABI, "Experimental Investigation of Creep Behavior of Bond Paper", Ph.D. Thesis, McGill University, Montreal, Canada (1978).
92. K. REZAI, "The Determination of Surface Displacements by Holographic-Electro-Optical Processing", Ph.D. Thesis, McGill University, Montreal, Canada (1981).
93. J. W. PROVAN and D. R. AXELRAD, "Probabilistic micromechanics of inhomogeneous solids", pp. 173–209, *Reviews on the Deformation Behavior of Materials*, Freund, Tel-Aviv, Israel (1977).
94. D. R. AXELRAD, J. W. PROVAN and S. EL HELBAWI, "Dislocation effects in elastic structured solids", *Arch. Mech. Stos.* 25, 801–810 (1973).
95. J. W. PROVAN and D. R. AXELRAD, "The effective elastic response of randomly oriented polycrystalline solids in tension", *Arch. Mech. Stos.* 28, 531–547 (1976).
96. J. W. PROVAN and O. A. BAMIRO, "Elastic grain-boundary responses in copper and aluminium", *Acta Metallurgica*, 25, 309–319 (1977).
97. W. BOLLMANN, *Crystal Defects and Crystalline Interfaces*, Springer, New York (1970).
98. O. A. BAMIRO, "On Elastic Grain Boundary Effects in Polycrystalline Solids", Ph.D. Thesis, McGill University, Montreal, Canada (1975).
99. D. R. AXELRAD and S. BASU, "Mechanical relaxation theory of fibrous structures", *Adv. in Mol. Rel. and Int. Proc.* 11, 165–190 (1977).

Bibliography 159

100. D. R. AXELRAD, "Theory of bond failure in hydrogen-bonded solids", *Adv. in Mol. Rel. and Int. Proc.* **15**, 51–69 (1979).
101. D. R. AXELRAD and S. BASU, "Mechanical relaxation of crystalline solids", *Adv. in Mol. Rel. Proc.* **6**, 185–199 (1973).
102. D. R. AXELRAD, "On the time-dependent behaviour of bond paper", in *General Constitutive Relations for Wood and Wood-based Materials* (ed. B. A. JAYNE, J. A. JOHNSON and R. W. PERKINS), Syracuse University, Syracuse, N.Y. (1978).
103. D. R. AXELRAD, S. BASU and D. ATACK, "Microrheology of cellulosic systems", in *7th Int. Congress on Rheology, Proc.* (ed. C. KLASEN and J. KUBÁT), Gothenburg, Sweden (1976).
104. D. R. AXELRAD, Y. M. HADDAD and D. ATACK, "Stochastic deformation theory of a two-dimensional fibrous network", in *11th Annual Meeting of Society of Eng. Science, Proc.* (ed. G. J. DVORAK), Durham, North Carolina (1974).
105. H. GHONEM and J. W. PROVAN, "Micromechanics theory of fatigue crack initiation and propagation", *Eng. Fracture Mechanics*, **13**, No. 4, 963–977 (1980).
106. J. W. PROVAN and C. C. I. MBANUGO, "Stochastic fatigue crack growth—An experimental study", *Res Mechanica* **2**, 53–72 (1981).
107. E. W. MONTROLL and B. J. WEST, "On an enriched collection of stochastic processes", in *Fluctuation Phenomena* (ed. E. W. MONTROLL and J. L. LEBOWITZ), North-Holland, Amsterdam (1979).
108. M. E. GURTIN, "Thermodynamics and the cohesive zone in fracture", *ZAMP* **30**, 991–1003 (1979).
109. C. C. I. MBANUGO, Ph.D. Thesis, McGill University, Montreal, Canada (1979).
110. H. W. HASKEY, "A general expression for the mean in a simple stochastic epidemic", *Biometrika* **44** (1957).
111. A. RÉNYI, *Wahrscheinlichkeitsrechnung*, Berlin (1962).
112. R. J. KNOPS and E. W. WILKES, "Theory of elastic stability", *Handbuch der Physik*, Bd. VI a/3, Springer Verlag (1964).
113. P. BILLINGSLEY, *Ergodic Theory and Information*, Wiley, New York (1965).
114. E. TONTI, *On the Formal Structure of Physical Theories*, Instituto Di Matematica del Politecnico di Milano, Milano, Italy (1975).
115. S. BASU, "On a General Deformation Theory of Structured Solids", Ph.D. Thesis, McGill University, Montreal, Canada (1975).
116. J. DIEUDONNÉ, "Recent developments in the theory of locally convex vector spaces", *Bull. Am. Math. Soc.* **59**, 495–512 (1953).
117. J. DIEUDONNÉ, *Foundations of Modern Analysis*, Academic Press, N.Y. (1960).
118. E. HILLE and R. S. PHILLIPS, *Functional Analysis and Semi-Groups*, Colloq. Publ. Amer. Math. Soc. (1957).
119. S. BOCHNER, *Harmonic Analysis and the Theory of Probability*, Berkeley, Los Angeles, University Press (1955).
120. D. G. KENDALL, "Some analytical properties of continuous stationary Markov transition functions", *Trans. Am. Math. Soc.* **78**, 529–540 (1955).
121. A. T. BHARUCHA-REID, *Elements of the Theory of Markov Processes and its Applications*, McGraw-Hill, New York (1960).
122. R. J. KNOPS and L. E. PAYNE, "Stability in linear elasticity", *Int. J. Solids Structures* **4**, 1233–1242 (1968).
123. M. F. Beatty, "Some static and dynamic implications of the general theory of elastic stability", *Arch. Rat. Mech. Anal.* **19**, 167–186 (1965).
124. R. S. Phillips, "Dissipative hyperbolic systems", *Trans. Amer. Math. Soc.* **86**, 109–173 (1953).
125. K. YOSIDA and S. KAKUTANI, "Operator-theoretical treatment of Markov processes and mean ergodic theorem", *Annals of Math.* **42**, No. 1 (1941).
126. K. YOSIDA, "The Markov process with a stable distribution", *Proc. Imp. Acad. Japan*, **16**(3), 43–48 (1940).
127. S. KAKUTANI, "Ergodic theorem and the Markov process with a stable distribution", *Proc. Imp. Acad. Japan*, **16**(3) 49–54 (1940).
128. K. YOSIDA, "Time dependent evolution equations in a locally convex space", *Math. Zeitschrift* **162**, 83–86 (1965).

Bibliography

129. S. MIZOHATA, *The Theory of Partial Differential Equations*, Cambridge University Press (1973).
130. A. PRÉKOPA, "On composed Poisson distribution", *Acta Math. Ac. Sci. Hungary* **3**, 317–325 (1952).
131. K. URBANIK, "Analytical characterization of some composed nonhomogeneous Poisson processes", *Studia Math.* **15**, 328–336 (1956).
132. J. W. PROVAN and H. GHONEM, "Probabilistic description of microstructural fatigue failure", *Proc. 2nd Symp. on Cont. Models of Discrete Systems*, pp. 407–430, University of Waterloo, No. 12 (1977).
133. D. R. AXELRAD, *On the Time-Dependent Behaviour of Bond Paper*, Nat. Sci. Found. Workshop, Syracuse University (1978).
134. D. R. AXELRAD, "Theory of hydrogen bond dissociation", *IXth Europhys. Symp. on Macromolecules*, Pol. Ac. Sci., Warsaw, Poland (1979).
135. J. T. ODEN, "Applied Functional Analysis", Prentice Hall Inc., N.Y. (1979).
136. D. R. AXELRAD, "Theory of deformation and stability of discrete media", invited lecture, in *Recent Developments in the Theory and Application of Generalized and Oriented Media* (ed. P. G. GLOCKNER, M. EPSTEIN and D. J. MALCOLM), Calgary, Canada (1979).
137. T. HOLLSTEIN, *Experimentelle Untersuchnugen zum Verhalten von Rissen bei Elasto-Plastischen Werkstoffverformungen*, Fraunhofer-Institut Für Werkstoffmechanik, Freiburg, West Germany, January (1982).
138. R. KANNAN, "Random operator equations", in *Dynamical systems* (ed. A. R. BEDNAREK and L. CESARI), Academic Press, N.Y. (1977).
139. P. L. CHOW, "Perturbation methods in stochastic wave propagation", *SIAM Review*, **17**, 57–81 (1975).
140. A. T. BHARUCHA-REID, "On the theory of random equations", in *Proc. of Symp. in Appl. Math.* (ed. R. BELLMAN) Vol. XVI (1964).
141. F. BROWDER, "Existence and uniqueness theorems for solutions of nonlinear boundary value problems", in *Proc. of Symp. in Appl. Math.* (ed. R. BELLMAN), Vol. XVI (1964).
142. G. ADOMIAN, "Stochastic Green's functions", in *Proc. of Symp. in Appl. Math.* (ed. R. BELLMAN), Vol. XVI (1964).
143. C. PRESTON, "Random fields", Lect. Notes in *Math.* No. 517, Springer (1975).
144. J. KAMPÉ DE FERIÉT, "Statistical fluid mechanics, two-dimensional gravity waves, in *Partial Differential Equations and Continuum Mechanics*, Madison, Wis., pp. 197–136 (1961).
145. J. B. KELLER, "Wave propagation in random media", in *Proc. Symp. Appl. Math.*, Vol. XIII, pp. 227–246, Amer. Math. Soc., Providence, Rhode Island (1962).
146. U. FRISCH, "Wave propagation in random media", in *Probabilistic Methods in Appl. Math*, Vol. I (ed. BHARUCHA-REID), Academic Press, New York (1968).
147. L. A. CHERNOV, *Wave Propagation in a Random Medium*, McGraw-Hill, New York (1960).
148. J. KAMPÉ DE FERIÉT, "Statistical mechanics of continuous media", *Proc. Symp. Appl. Math.* **13**, 165–198, Am. Math. Soc., Providence, Rhode Island (1962).
149. B. J. USCINSKI, *The Elements of Wave Propagation in Random Media*, McGraw-Hill, New York (1977).
150. D. R. AXELRAD and M. OSTOJA-STARZEWSKI, "Probabilistic theory of elastic longitudinal wave propagation in discrete media", *9th U.S. National Congress of Applied Mechanics*, Cornell University, Ithaca, N.Y. (1982).
151. N. METROPOLIS and S. ULAM, *Amer. Math. Monthly*, **50**, 252 (1953).
152. M. OSTOJA-STARZEWSKI, "Microdynamics of Structured Solids", Ph.D. Thesis, McGill University, Montreal, Canada (1983).
153. R. SYSKI, "Markov chains and probability potentials", in *Probabilistic Methods in Appl. Math.* (ed. A. T. BHARUCHA-REID), Academic Press, New York (1973).
154. P. A. MEYER, *Probability and Potentials*, Ginn (Blaisdell), Boston, Massachusetts (1966).
155. I. R. MCDONALD, "Molecular dynamics, principles and applications", in *Microscopic Structure and Dynamics of Liquids*, Plenum Press, New York (1978).
156. J. P. BOON and S. YIP, *Molecular Hydrodynamics*, McGraw-Hill, Adv. Series, New York (1980).

Bibliography 161

157. S. W. Lovesey, *Dynamics of Solids and Liquids by Neutron Scattering*, Springer, Heidelberg (1977).
158. J. C. Dore, G. Walford and D. I. Page, *Mol. Physics* **29**, 565 (1975).
159. J.-P. Hansen, "Correlation functions and their relationship with experiments", in *Microscopic Structure and Dynamics of Liquids* (ed. J. Dupuy and J. J. Dianoux), Plenum Press, New York (1977).
160. P. A. Egelstaff, *An Introduction to the Liquid State*, Academic Press, New York (1967).
161. M. Metropolis, A. W. Rosenbluth, M. N. Rosenbluth, A. N. Teller and E. Teller, *J. Chem. Phys.* **21**, 1087 (1953).
162. A. Bellemans, "Computer simulation of molecular liquids", *Adv. Mol. Rel. Int. Proc.* **24**, No. 2, 107–114 (1982).
163. P. C. Martin, *Measurements and Time Correlation Functions*, Gordon and Breach, New York (1968).
164. P. C. Martin and S. Yip, *Phys. Rev.* **170**, 151 (1968).
165. C.-H. Chung and S. Yip, *Phys. Rev.* **182**, 323 (1969).
166. R. Kubo, *J. Phys. Soc. Japan* **12**, 570 (1957).
167. R. Kubo, *Statistical Mechanics*, North-Holland, Amsterdam (1965).
168. M. S. Green, *J. Chem. Phys.* **20**, 1281 (1952).
169. R. Zwanzig, in *Statistical Mechanics* (ed. S. A. Rice, K. F. Freed and J. C. Light), University of California Press, Chicago (1972).
170. L. P. Kadanoff and P. C. Martin, *Ann. Phys.* **24**, 419 (1963).
171. M. C. Wang and G. E. Uhlenbeck, *Rev. Mod. Phys.* **17**, 323 (1945).
172. P. N. Pusey, "Intensity fluctuations spectroscopy of charged Brownian particles: the coherent scattering function", *J. Phys. A: Math. Gen.* **11**, No. 1, 119–135 (1978).
173. B. J. Ackerson, *J. Chem. Phys.* **64**, 242–246 (1976).
174. B. J. Berne, in *Photon Correlation Spectroscopy and Velocimetry* (ed. H. Z. Cummins and E. R. Pike), Plenum, New York (1977).
175. G. I. Taylor, *Proc. Roy. Soc. A*, **102**, 180–189 (1922) and **104**, 213–218 (1923).
176. R. E. Meyer, *Introduction to Mathematical Fluid Dynamics*, Wiley-Interscience, New York (1971).
177. R. Abraham and J. E. Marsden, *Foundations of Mechanics*, Benjamin, London (1978).
178. G. K. Batchelor, "Developments in microhydrodynamics", *XIVth Int. Congress I.U.T.A.M.*, North-Holland, Amsterdam, pp. 33–35 (1976).
179. D. R. Axelrad, "Micromechanics of fluids", in *Mechanics of Structured Media* (ed. A. P. S. Selvadurai), Carleton University, Ottawa, Canada (1981).
180. D. R. Axelrad, "Markov theory in the mechanics of discrete fluids", *Int. J. Eng. Sci.* **20**, No. 2, 181–186 (1982).
181. E. G. D. Cohen, private communication (1980).
182. C. A. Croxton, *Statistical Mechanics of the Liquid State*, Wiley, Chichester (1980).
183. C. J. Preston, *Gibbs States on Countable Sets*, Cambridge Tracts in Mathematics, Cambridge Univ. Press, No. 68 (1974).
184. F. Spitzer, "Markov random fields and Gibbs ensembles", *Am. Math. Monthly*, **78**, 142–154 (1971).
185. R. Holley, "Markovian interaction processes with finite range interactions", *The Annals of Math. Statistics* **43**, No. 6, 1961–1976 (1971).
186. N. T. J. Bailey, *The Mathematical Theory of Epidemics*, Hafner Publ. Co., New York (1953).

Subject Index

Abstract Cauchy problem 18
Abstract dynamical system 59, 70, 118
After-effect function 134
Auto correlation 39, 130
  function 137
Auto covariance function 39
Automorphism 17, 54

Basic stochastic processes 33
Boltzmann constant 123
Bonding in fibrous structures 87
  probabilistic model of 91
Boundary value problems 110
  Dirichlet 110
  Neumann 111
  probabilistic form of the
    Dirichlet problem 112
    Neumann problem 113
Brownian motion 47, 135

Central force model 71
Chapman–Kolmogorov relation 46, 55
Characteristic
  frequency 132
  function 27
  length 148
  wavenumber 132
Characteristics of stochastic functions 35
  second order 38
Chebyshev inequality 39
Collision time 139
Compactness 7
Computer simulation 85, 157
  techniques 132
Configuration 63
Configurational integral 125
Connectedness 7
Contraction group of C_0-class 18
Covariance functions 38, 136
Covering 6
Creep and relaxation 87
Cross-variance 39
Cumulant functions 39

Deformation kinematics 93, 94
Density regimes 147

Diffusion 121
Diffusion constant 139
Dispersion time 115
Dissipation effect 135
Dissociation process in fibrous structures 90, 92
Dynamic structure factor 131
Dynamics of structured solids 114

Elements of the structure 59
Ensemble
  canonical 151
  Gibbs 63
Equilibrium probability density 124
Equilibrium time correlation function 129
Ergodic
  class of sets 152
  system 130
  theorem 49
Eulerian representation 146
Event 19, 65, 98
External field 134

Failure of polycrystalline solids 90
Field variables (quantities) 54
  definitions of 62
Fluctuation-dissipation theorem 135, 138
Fluctuations 121
Fokker-Planck equation 139
  multidimensional 142
Fracture and bond failure 87
Friction constant 136
Function(s)
  Borel 7
  distance 7, 77
  integrable 13
  mean and correlation 38
  measurable 13
  pair-correlation 129
  stationary random 34
Functional
  continuous 14
  correlation 43
  linear 14
  Minkowski 10
  random 42

164 Subject Index

Fundamental concepts of probabilistic mechanics 62

General probabilistic deformation theory 93
Generalized microdeformation 78, 80, 95
Generalized microstress 78
Generalized random field 119
Grain boundary 74, 85
  effects in 74
Groups 4
  Abelian 4

Hamiltonian 122
  unperturbed 133
Hard-sphere model 72
Higher moments 29
Homeomorphism 7
Hydrodynamics 121

Infinum 12
Infinitesimal
  generator 17, 18, 56
  volume 124
Interaction
  effects 70
  shortrange 74
  potential 71, 72
  zone 85
Interatomic and intermolecular forces 70
Internal
  energy 124
  interfaces 59
  time 115

Kinematics of simple fluids 145
Kinetic energy 123
Kolmogorov equation
  backward 140
  forward 141

Lagrangian representation 146
Limit of the linear bounded operator
  strong 17
  weak 18
Linear response theory 126
Local dynamical variables 127
Long wavelength 121

Macrodomain 63

Macroscopic material body 59
  time 115
Mapping(s) 2
  bijective 3
  image of 3
  injective 3
  inverse of 3
  surjective 3
Markov
  chain 50
  homogeneous 51
  interactive processes 153
  process 46
  in probabilistic mechanics 46
  with stable distribution 50
  property 47
  random field 46, 57, 154
  time 52
  theory for discrete fluids 148
Markovian
  deformation process 80
  model of crack propagation 88
  fracture 88
Mass density 64
  in fluid 143
Material characteristics 85
Material operator 77
  evolution of 81
  inverse of 78, 80
Material point 63
Measure(s) 11
  Borel 13
  bounded 12
  complete 12
  probability 20, 68
  product of 12
  regular 12
Mechanical states of medium 61
Mesodomain 63
  configurational 143
Microelement 63
  in fluid 142
Microstates 123
Microstress distribution 80
Microstructure 59
  of solids 82, 83
Molecule(s) 72
  configuration of 142
  diatomic 72
Molecular dynamics 142
  of simple fluids 122
Moment generating function 30
Momentum and energy density gradients 121
Monte Carlo method 114, 151

Norm 12, 97

Subject Index 165

Observables 60
  expected value of 61
Operator(s)
  bounded linear 16
  boundedness and continuity of 16
  linear 14, 15
  Liouville 127
  material 77, 110
  mean-square differential 37
  on Banach space 77
  one-parameter 8
  projection 15
  self-adjoint 16
  semi-group of 18
  two-parameter 117
  transition 56
Operational constitutive relations 79
Ornstein-Zernicke relation 131
Outcome 19

Pair Potential 73
Passage time 115
Phase points 123
Planck's constant 124
Potential
  interaction 71, 72
  Lennard-Jones 74
  modified Morse 74, 84
Power flux 119
Power spectrum
Probability
  conditional 21
  conditional probability space 26
  definitions of 19
  measure 21
    conditional 54
Probability density function 24, 124
Probability distribution function 24
  Gaussian 29
  joint
  Maxwell 139, 125
  n-particle 126
  Poisson 29
  radial 130
  single particle 126
Probabilistic function space
  mechanics of fluids 122
  structure of 66
Process
  continuous time-parameter 49
  discrete time-parameter 49
  Gaussian 40, 136
  Gaussian Markov 136
  Markov 46
  Poisson 41, 150
  random deformation 54

temporally homogeneous 56
Projection 15
Propositional logic 60

Quantum statistical mechanics 124
  stochastic mechanics 60
Quasi-norm 10

Random endomorphism 119
Random experiment 19
Random field 42
  generalized 44
  Gibbs 146
  Markov 46
  second order Gaussian 44
Random fluctuating force 135
Random propagation velocity 115
Random variable 23
  Banach space-valued 23
  continuous 23
  definition of 23
  discrete valued 23
  expectation of 28
  independent 31
Random vectors 25
Real fluid 147
Recurrence relation 152
Relative distance 129
Response theory formalism 132
Ring 4

Scalar
  product 11
  field 14
Scattering function 131
Semi-group
  C_0-class 18
  dynamical 47, 53
  operator 8, 18, 117
  property 56
Semi-norm 9, 97
Semi-ring 4
Separation distance 72
Sets(s)
  Baire 13
  Borel 5
  closure of 5
  definitions 2
  difference 2
  disjoint 12
  equivalence class of 3
  interior of 5
  intersection 2
  partition of 2

Sets(s) (*Contd.*)
　relations　3
　subset of　2
　system of　4
　union of　2
　window　66
σ-algebra　4, 20, 60
　Borel　5
σ-field of events　20
Space(s)
　adjoint　76
　Banach　10
　duality of　75
　Euclidean R^n　42
　Frechet　45
　function　65
　Hausdorff　6
　Hilbert　11
　linear　8
　locally convex　8
　quasi-normed　10
　metric　7
　probability　1, 11, 20
　product of　6
　sample　19
　Schwartz　45
　separable　8
　topological　5
Spectroscopic studies　85
Spherically symmetric interactions　122
Stability of microstructure　102, 106, 109
State vector
　in fluids　67
　in solids　67
Statistical
　average　123
　state　60
　states of elements　60
Stochastic force autocorrelation function　138

Langevin type differential equation　135, 137
Stress distribution, evolution of　80
Structure factor　129
Super-harmonicity　120
　martingale　120
Supremum　12

Temperature　123
Thermodynamic equilibrium　122
Transition densities　46
　intensities　90
　probability　47, 50
　n-step　152
Transmission
　coefficient　114
　operator　118
Transport
　characteristics　121
　coefficients　135
　phenomena　135
Trial　19
Triplet　11

Unidirectional wave propagation in solids　119
Unit cell　86

Virial function　123
Viscous flow　121

Wave propagation　114
Wave evolution　115
Wave velocity vector　115
Wavefront evolution　120
Weak topology　77